THE
APOTHECARY'S
WIFE

KAREN BLOOM GEVIRTZ spent nearly three decades as a professor of English at American universities, most recently at Seton Hall University. Gevirtz earned a BA in English at Brown University, where she was also pre-med and a research assistant in a neurochemistry lab. She has a PhD in British Literature and has received fellowships and grants for her archival research. Internationally recognized for her scholarship on women and writing in the seventeenth and eighteenth centuries, she has authored academic articles, chapters and several scholarly books, and co-edited a collection of essays. She lives in New Jersey, USA.

KAREN BLOOM GEVIRTZ

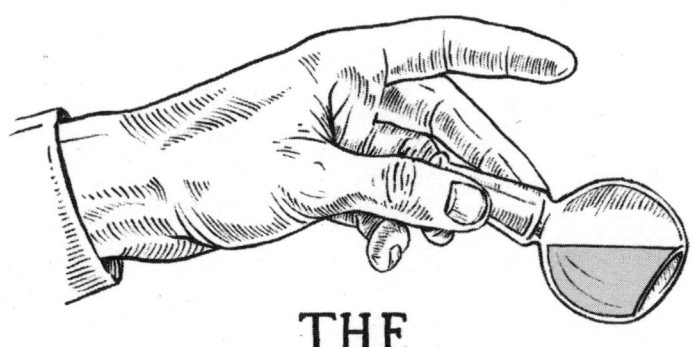

THE
APOTHECARY'S
WIFE

The Hidden History of Medicine and
How it Became a Commodity

HEAD
ZEUS

An Apollo Book

First published in the UK in 2024 by Head of Zeus Ltd,
part of Bloomsbury Publishing Plc

9 7 5 3 1 2 4 6 8

A catalogue record for this book is available from the British Library.

ISBN (HB): 9781803286990
ISBN (E): 9781803286976

Typeset by DivAddict Typesetting Solutions Ltd

Printed and bound in Great Britain by
CPI Group (UK) Ltd, Croydon CR0 4YY

Head of Zeus Ltd
First Floor East
5–8 Hardwick Street
London EC1R 4RG
WWW.HEADOFZEUS.COM

For Jessica,

who said, "What about Chawton?"

Contents

Introduction

This book began in a 400-year-old basement. The archivist of Chawton House Library said to me, "And of course, you'll want to see our copy of Blackwell's *Herbal*," and I replied, "Yes." I was not exactly being honest. I had travelled 3,000 miles to research how eighteenth-century women incorporated the Scientific Revolution into their daily lives, and "Blackwell's *Herbal*" was not on my to-do list. I did not even know what it was. But when a librarian or archivist suggests that you take a look at something, you say yes – and that was when the book that I thought I was writing ended and this one began.

"Blackwell's *Herbal*" was shorthand for *A Curious Herbal, Containing Five Hundred Cuts, of the Most Useful Plants Which Are Now Used in the Practice of Physick*, created by Elizabeth Blackwell between 1735 and 1739. A herbal is a compendium of medicinal plants made to be used and referenced; however, Blackwell made hers for show. It consists of 500 etched and hand-coloured pictures, 125 etched pages of information, and another dozen etched pages of front and back matter, including dedications and indexes in English and Latin. It is beautiful, in the same family of works as John James Audubon's *Birds of America* (1827) and Maria Sibylla Merian's *Metamorphosis Insectorum Surinamensium* (*Metamorphosis of the Insects of Suriname* [1705]). *A Curious Herbal* was also the first printed herbal by a woman that I had seen or heard of. Aside from being a remarkable work, two things about it struck me at that first reading. One was the number of eminent physicians, apothecaries, and Fellows of the Royal Society who had supported and encouraged Blackwell. They included Sir Hans Sloane, president of the Royal College of Physicians (RCP); Richard Mead, a leading physician of the early eighteenth century; Isaac Rand, prae-centor of the apothecaries' guild teaching garden in Chelsea; and James Douglas, premier anatomist and man-midwife (obstetrician). It was very odd that Elizabeth Blackwell, an undistinguished Scottish woman married to a man arrested for bankruptcy, was connected with such an illustrious group.

The second thing that struck me as I read the *Curious Herbal* was this line: "The Herb Women sell the Leaves of the *Helleboraster*, or Bear's-foot, or *Sphondylium*, or Cow parsnep [*sic*], instead of this Plant, to those that are ignorant." It was the first mention of women other than as patient and to my disappointment, it was not sympathetic. It also had nothing to do with medication per se; Blackwell was accusing the "Herb Women" of an economic crime. Similar passages occurred peri-odically throughout the two volumes, all levelling the same accusation that the herb women were guilty of fraud and of preventing the "ignorant" from buying the proper herbaceous material from those who were honest and knowledgeable. It seemed absurd to me that anyone in 1735 would think that herb women competed with apothecaries and physicians in any way, but there it was. Blackwell's patrons were illustrious apothecaries and physicians; whatever she wrote had to be pleasing to them, perhaps articulating their views. Why was this group so threatened by these women? Why did they consider the threat to be economic?

The fact that I was asking those questions meant that there was something very wrong with the standard history of medicine, what I might call "the Triumph of Modern Medicine". That (his)story proceeds roughly like this: Once upon a time, ancient Greeks and Romans such as Hippocrates and Galen began to establish medicine as a field of knowledge and a form of practice, but everything remained crude until the Scientific Revolution, which brought in the scientific method, true knowledge, and the wonderful advances that continue to this day and from which humanity benefits, the end. One of this story's subplots is the triumph of modern medication: Once upon a time, medication was revolting and harmful because it was made by ignorant crones, whereas one of the wonderful advances of the Scientific Revolution was effective medication, which is what we have now, the end. There is nothing in either of these accounts of the past that has to do with money. These are "triumph" or "rise" histories, in which a revolution introduces something entirely new. Competition meant that "new" did not necessarily mean "better" to people in the 1730s.

Furthermore, A *Curious Herbal* showed that the medication being prescribed by physicians and made by apothecaries used the same organic materials as those used by ordinary people. The title page announced that the book depicted 500 "of the most useful Plants which are now used in the Practice of Physick". In case this was not clear enough, each entry about a plant informed readers why it would be prescribed and how it would be prepared, often citing renowned authorities in medical botany. On rare occasions, Blackwell relied solely on what "common folk" used a plant for; occasionally she cited both professional and common usage. It appeared that professionals and housewives alike used herbal ingredients for their medications. Could "modern" medicine have been perceived as superior? Yes, of course. Could it have been actually superior? It did not look like it. This led to another and particularly bothersome question. If the medications were the same, why did people stop getting treatment at home and start getting it from physicians and apothecaries? If money was an issue, as Blackwell's book suggested, how did the professionals get people to start paying for something that for hundreds of years they had been getting for free?

That is the central conundrum of this book. For centuries, ailing people got medication and treatment at home. Their medication was made by a member or a friend of the family, it was made from time-tested recipes and ingredients, and it was made right there. Furthermore, there was no charge. Then, ailing people started getting medication outside the home, travelling while ill to physicians or apothecaries. The medication that the patient purchased was made by a stranger from unknown ingredients according to an unknown recipe in a room that the patient was not permitted to enter. *The Apothecary's Wife* explains how, when, and through what means the Scientific Revolution catalysed the replacement of medications made at home with medications made by apothecaries and prescribed by physicians. This book also reveals that the Scientific Revolution transformed the entire concept of medication from a household item into a commodity, something to be sold and bought. An entire healthcare system – women's domestic medicine – was switched out for the system we have now – that is, for-profit medicine. There is nothing about the current medication system that is natural or eternal or inevitable. What exists was made by people at a certain time making certain decisions and doing certain things. We pay for medicine because we think of medicine as something to be paid for. We did not always think so. We choose whether to continue.

Therefore, *The Apothecary's Wife* is also something of an economic history. I explain in the coming pages that the Scientific Revolution and what became science and modern medicine have always been entwined with issues of personal and public gain, and with profit as well as benefit. Consequently, this book is not only a history of medicine but also a history of the economic system developed for the circulation of medication as a commodity. Such an economic system protects, normalizes, and obscures medication's commodity status. Powerful, dangerous aspects of it, such as unequal access to medication, the regulation of drugs but not of prescriptions, and the embracing of highly technical language (not to mention the fostering of educational curricula that do not prepare people to deal with that language) were knowingly built into it during its construction. These issues and the current global for-profit system are neither inevitable nor immutable. For-profit medication was created by choices; consequently, it can be retained, modified, or eliminated by choices.

This book is built on a tremendous body of material, only a small portion of which is visible in the reading. I have drawn on texts created by both men and women, including correspondence, diaries, recipe books, household budgets, personal accounts, plays, poems, sermons, narrative fiction, paintings, prints, tracts, treatises, textbooks, scientific articles, newspapers, advertisements, gardening handbooks, housekeeping manuals, herbals, pharmacopoeias, medical lectures, students' notes, annotations, inventories, prescriptions, municipal and parish records, guild archives, judicial papers, royal declarations, and legislation, both national and local. Although each chapter is built around two central figures, in fact readers will meet a whole range of people who lived and contributed to the transition. They will also hear the voices of long-dead people, including women. For millennia, women's lives were not well documented and as a result, finding them can be difficult. "When it comes to the lives of the other half of humanity," Caroline Criado Perez wrote, "there is often nothing but silence." The hundreds of surviving recipe books, fortunately, give them voice. These books speak of women's lives and worries – the ailments they treated, the meals they cooked, the children they bore or miscarried, and how they and their families grew and flourished or suffered. Every ailment afflicted a human being and concerned not only the sufferer but also the people who loved them and cared for them. Who to consult and where to get treatment could be life-and-death decisions; these were not academic issues for them any more than they are in this century. An account of such a watershed moment should include everyone who made that moment happen. This account does.

This book is not many things. It is not a book about how men are evil because they suppressed and oppressed women. Lived life is more complex than this simple binary and the historical record bears that out. Men and women participated in the Scientific Revolution, and men and women resisted it. Elizabeth Blackwell was not the innocent, unwitting stooge of Machiavellian physicians and apothecaries; *A Curious Herbal* was her idea and she genuinely advocated for what is sometimes called the New Science. *The Apothecary's Wife* is also not a diatribe against capitalism. For one thing, that would be hypocritical considering that this book is also a commodity. Instead, it is a critical analysis of one aspect of the development of Western capitalism, which itself has more than one strain. This book is not anti-science, anti-medication, or anti-"modern" medications. It is not advocating homeopathy or "natural" medications or supplements. Actually, it is not advocating anything. *The Apothecary's Wife* offers knowledge and insight about what happened in the past, and how and why it happened, so readers can better understand the present. The book is mine, the history is shared, and the choices are yours.

A few preparatory notes are in order. One concerns the term "Scientific Revolution". Scholars do not use it any more, having established in the last few decades that "science" in its current meaning did not exist until the start of the nineteenth century, and that there were numerous strands of innovative thought, not the single one implied by the singular "revolution". It is a convenient term precisely because it is so familiar and generally known, however, so I encourage readers to remember that it is used in a capacious sense. Another note concerns the idiosyncratic spelling and punctuation characteristic of the seventeenth and eighteenth centuries. With a very few, recognizable exceptions, texts are quoted in their original form. If a quotation appears to be gibberish, I recommend reading it aloud because sound often reveals the word when the eyes cannot make sense of it. In places where spelling might appear to be a typo, I have written [*sic*]. In places where writing was illegible, I have written [*illeg*]. Bad handwriting transcends place and time, and different periods in history also favoured different handwriting styles, some easier to read than others. There was a significant change in the type of handwriting at the end of the seventeenth century, which can be helpful when dating a document or recognizing the age of a writer. In some recipe books, for instance, entries start with one handwriting style and continue through at least one or two others, while in others, an occasional recipe from an older contributor appears among those written by younger women. Ordinary wear and tear also have their impact. One eighteenth-century recipe book I read had a charred hole through the middle of several consecutive pages. Not hard to guess what kitchen mishap caused that.

Place names posed another challenge. Before 1603, England (which controlled Wales and Ireland) and Scotland were separate nations ruled by different monarchs: the Tudors in England and the Stuarts in Scotland. In 1603, Queen Elizabeth I of England, a Tudor, died childless and her cousin, King James VI of Scotland, a Stuart, was crowned King James I of England. Between 1603 and 1707, England and Scotland were different nations governed by different parliaments, but ruled by the same monarch. In 1707, the two countries united "into One Kingdom by the Name of Great Britain". In this book, the term "British Isles" refers to the agglomeration of England, Scotland, Wales, and Ireland before 1707, while "Great Britain" and "Britain" refer to it after 1707. "Great Britain" specifically covers England, Wales, and Scotland. The term "United Kingdom" appears only in Part Two. I excluded Ireland from my research because, unlike the other countries, it was actively and deliberately exploited as a colony during this period and consequently has a very different history.

Rather than footnote every unusual word, which would be intrusive and cumbersome, I have included a list of frequently used medical and apothecary terms. These definitions are taken from the *Oxford English Dictionary*, medical dictionaries of the time, and explanations by contemporary medical writers.

ana	abbreviation for "of each" in a recipe
antiscorbutic	medication for treating scurvy
bolus	medicine in a round or globe shape
cataplasm	a poultice
cathartic	a purgative that works on the bowels
clyster or glister	an enema
compound	medication made of more than one ingredient
costive	constipated
decoction	a liquid medication made by boiling a small amount of medicinal substance in a lot of water
drachm, dram	a very small quantity of medication (apothecary measure)
electuary	medication made by mixing medicinal powder with honey or another sweet, sticky substance
emplastrum	medicinal paste or salve smeared on; sometimes used for "plaister" or "plaster"
flux	a substance flowing out of the body, including menses
fomentation	warm soft material soaked in medicinal liquid and applied to the body
jalap, julep	water with sweet syrup and medicinal substance dissolved in it
lithontriptic(k)	medicine that breaks up bladder and kidney stones
mead	a drink made with fermented grains and honey
meathe	spiced mead
menstruum	a solvent

officinal   official stock preparation made by apothecaries and prescribed by physicians

pectoral   medicine good for digestion or as an expectorant

plaister, plaster   cloth soaked in medicinal liquid and dried, to be laid on the afflicted area so that medication will sink in through the skin (also emplaister, emplaster)

scruple   one-third dram; apothecary measure

simple   a medicine made of only one ingredient

vulnerary   medicine for treating wounds

Part One

I

Kitchen Physic
Is the Best Physic

T he scene: a richly furnished bedroom in England. Thick
tapestries on the walls decorate and insulate, the latter
function being especially important because there is no
glass in the windows, which are themselves narrow and not
very high. Hangings, possibly embroidered, on all four sides
of the wooden bed and a canopy on top provide warmth and
privacy. The bed enclosed by those fabric walls is occupied by
a wealthy, powerful person who is unwell. Worse, the year is
somewhere around 1250 CE. On the positive side, this ailing
noble of medieval England has a physician on staff in the household. It is time for said physician to earn his salary and keep.

Summoned to the bedside, the physician asks a lot of questions: When did you start to feel ill? Can you describe your
symptoms? What have you been eating and drinking? What
colour is your urine? How many times have you had a bowel
movement in the last twenty-four hours? At what date and
time were you born? He also checks the patient's pulse, the
colour of the tongue and the whites of the eyes, and pal-
pates any body parts causing distress. Having collected the

necessary information, the physician gets to work. He draws up his patient's horoscope because the sign and planet under which someone is born will have a significant influence on their treatment. He might have an almanac handy for reference. He consults tomes in Greek or Latin. Then he makes a diagnosis and gets to work. His remedy probably includes a procedure like bloodletting, applying leeches, or cupping. It will involve a medication, which he will make himself.

Even the thirteenth-century English patient knows that illness has to do with an imbalance of humours. The physician's job is to determine what humours are unbalanced, why they are unbalanced, and how to recalibrate them. This model of the body and role of the physician came from ancient Roman and Greek philosophers, primarily Hippocrates (c. 460 – c. 375 BCE), after whom the physician's oath is named, and Galen (129–210 CE). According to the Galenic model, the body has four humours and four elements; people are healthy when the humours and elements are in balance, and unwell when they are not. The four humours correspond to fluids in the body – black bile, yellow bile, blood, and phlegm – and to qualities: cold and dry, hot and dry, hot and moist, and cold and moist, respectively. Different humours predominate depending on gender and age. Young men are hot, old men are cold; too much phlegm in a soldier makes him a coward, too much blood makes him reckless. Medication helps the body reset its humours, perhaps by putting the body through a physical ordeal like vomiting or sweating, but sometimes in ways the patient cannot feel. Neither one drug nor one dose is likely to fix everything necessary. The physician – or at least his medications – will be needed by our ailing aristocrat for a while.

As it has been for centuries, medication is made from materia medica, which is roughly translated as the materials of medicine. The more unusual ingredients include spices, animal parts, bits of mummies, metals, gems, pearls, coral, ambergris, and dung. However, the vast majority of materia medica is herbaceous, and there are many medicinal plants to choose from. At least one of the physician's books in Latin is a herbal, an encyclopaedia of medicinal plants that explains what to use them for and how. All herbals, even in the third millennium, are descendants of *De Materia Medica*, compiled by a first-century Greek physician in the Roman army named Dioscorides.

As it is the middle of the thirteenth century and the only places that use printed material are Korea and China, the physician's herbal is written by hand. If it is illustrated, chances are that the illustrations are beautiful, possibly witty, but unhelpful in identifying the plants. This is where training and the local shops can be of help. The shops are owned and run by apothecaries, a fairly new occupation for the time. In the middle of the thirteenth century, apothecaries belong to the Guild of Pepperers, which imports and sell spices, most notably and obviously pepper. Generally, pepperers sell their goods wholesale and in bulk, while apothecaries are customer facing. This is not a hard and fast distinction, however, and a pepperer could also have an apothecary business. Apothecaries sell exotics like spices, which are among the most expensive substances in Western Europe, as well as wine and herbs. Some apothecaries also sell medicines. Anyone, including physicians, can get hard-to-find ingredients like cinnamon and nutmeg at the right apothecary. But to suggest that the apothecary himself has something valuable to contribute medically or medicinally would elicit laughter. You might as well ask the local wool merchant for healthcare advice.

Physicians serve a very small group; then as now, not many people are wealthy, let alone very wealthy. Thirteenth-century

London is an urban centre, but it has fewer than 80,000 inhabitants, a small percentage of England's population. Most sick people get their treatment at home from the trusted, expert female head of the family. Making medication is just one of her domestic responsibilities. Women run the household, which means supervising and doing tasks like cooking, cleaning, sewing, weaving, mending, repairing, carrying and delivering and nursing and raising children, washing laundry, making household goods like candles, carding wool, spinning thread, preserving food, making wine and brewing beer, and training children and servants. They also work in the fields or in the shop, look after farm animals like chickens and pigs, go to market to buy and sell goods, maintain the kitchen garden (which has vegetables and medicinal plants), and gather ingredients from around the countryside. Domestic space is a capacious domain ranging from the garden to the barn to the kitchen to the parlour.

The older women in the family teach the next generation of women the knowledge and skills to diagnose illness, make medication, and administer it. A housewife's remedies are highly personal: they have been used on people who were near and dear, they have been seen (or thought) to work at least to some degree, and they do not appear to have hurt anyone. Like the physician, the woman of the household uses herbaceous materia medica; unlike him, she would never use arsenic, mercury, ground emerald, bloodletting, cupping, and the rest of it. She does not have a mortality rate, either. Nor does she charge a fee. Women are trusted for a reason. As one popular saying put it: "Kitchen physic is the best physic." A woman with a particular talent for healing can serve a whole neighbourhood in exchange for coins, goods, or services. She is not likely to be accused of witchcraft – accusations of witchcraft have more to do with interpersonal conflict and anxiety about political, social, or religious instability than with the status of an individual in a community. Besides, good healers are hard to find. All in all, when it comes to medicine, it is better to be a sick farmer than a sick earl. The farmer is more likely to get relief and less likely to suffer or die from the medication provided at home by women than the earl is from the physician's professional treatment.

But suppose – since we are imagining – that a tear in the fabric of space–time allows our afflicted sufferer to cross into the next century in search of better treatment. Unfortunately for the ailing time traveller, summoning the physician in this century results in the same experience as it did in the last: he asks the same questions, does the same physical examination (if any), checks the pulse, draws up a horoscope, and spouts a lot of Latin and Greek before making a diagnosis and issuing orders for treatment. He also uses the same techniques: bloodletting, leeches, cupping, and medication made from materia medica. This remarkable consistency is fundamental to the system and philosophy of education at this time. The fourteenth-century physician is a product of scholasticism, the philosophy that all necessary medical knowledge is lodged with ancient writers, and that learning their insights is the only way to become a capable practitioner. Scholastics have great faith in the ancient authorities; if Galen said it, it must be so. Medical students attend lectures where they listen to someone read aloud books by ancient Greek and Roman authorities and frantically write down as much as possible verbatim.

In fourteenth-century England, not much has changed for women either. Another century has passed where they serve as the trusted, reliable medical authority for the majority. Although, like everyone else, the woman of the family believes that the body conforms to Galenic theory, she does not believe that whatever the ancients wrote down was correct or that knowledge and expertise depend on book learning. She has been trained at home and that education serves her people well. She makes medications in her kitchen using recipes that are handed down and time-tested. Never mind what Galen said – if his advice or anything else proves fatal, it is not used again.

That is one reason why domestic medication is more effective than whatever physicians are administering. Another reason is how women and physicians process their ingredients. Both use simples, which are liquids made from one ingredient like cloves or marjoram, and compounds, which are combinations of simples. Some medications are simples, some are compounds, and some are combinations of compounds. Physicians use simples far less than women do and they process their ingredients far more. Both groups, for example, recognize liquorice root's usefulness in treating several forms of respiratory ailments, an effect confirmed by twentieth-century laboratory analysis. How they use it, on the other hand, is quite different.

Suppose our ailing English patient has a respiratory disorder. A housewife would wash a handful of Hope leaves (probably hops) in "running fayre water", add a handful of pitted "great raysins", a "good stick of licguoris [liquorice] well bruised", and "a little orgomye" (probably agrimony); soak it all together; strain the liquid; add sugar; and have the patient drink some twice a day. The physician would transform the liquorice into a powder or pill or oil and combine it with any number of other ingredients similarly transformed. The wealthy time traveller's physician's order would look like this:

℞. of the seds of white poppie. ℨ.ss. gumme arabick, Amylum, and gumme tragacant. ana. ℨ.j, ss. seedes of cucumbers, citrons gourds, melons, quinces. ana. ʒ.iij. burnt Ivory, iuice of licorace. ana. ʒ.j.ss. penidies, as much in weight as all the rest. Make a pouder, of the which minister daily everie morning. ʒ.ij with sirup of poppy or roses.

Translated, the prescription combines half a dram of white poppy seeds; five drams halved (two and a half drams) each of gum Arabic, plant starch, and gum tragacanth; three drams each of seeds from cucumbers, citrons gourds, melons, and quince; five drams halved of burned ivory and liquorice juice; and the equivalent weight of this mixture in pulvis (a medicinal powder). After grinding it all up together, four drams of the new compound powder are to be mixed with syrup of poppy or syrup of roses and licked up every morning. By the time the liquorice gets into the patient, it is so changed and so small a part of the whole that any medicinal qualities are greatly reduced, if not eliminated.

All illness takes second place to the Black Death, which strikes Europe in the middle of the fourteenth century. It begins in 1347 when a fleet of ships from Asia bearing plague victims arrives in Messina, Sicily, and shortly after, when another ship arrives in Venice. The plague quickly overwhelms the continent, kills one-third of the population, and lays human nature bare. Of particular concern to our sick time traveller is the response of physicians, which is disappointing at best. As a group, they are notorious for fleeing or refusing to treat plague sufferers. (They did not wear the famous beaked plague mask. That would come later.) Physicians are not unique in this regard but considering their profession, it is particularly egregious. Their descendants would do it again at the next major outbreak of plague 300 years later.

The Black Death is a turning point for the apothecaries. Until then, they had been evolving slowly into something distinct from the pepperers. The plague speeds up that process, and the physicians' failure to treat patients makes people turn elsewhere for prophylactics, cures, and treatments. Apothecaries are a logical choice because they sell herbs, spices, and other elements of the materia medica. Some apothecaries have already started to offer medical advice. After all, these shopkeepers have been selling their wares to physicians for some time, so they have knowledge of what the different herbs and spices can do.

The 1340s are no time to linger in, so imagine that the sick English aristocrat leaps forward another hundred years or so, to the middle of the fifteenth century. Looking out of the window, it is clear that London has recovered literally and demographically from the Great Plague: more than 100,000 inhabitants have stretched the city beyond its original walls. Henry VI and Margaret of Anjou sit on the throne, but not for long; Joan of Arc and the French armies have just finished drubbing the English; William Caxton has not yet imported his printing press; Martin Luther has not yet nailed his *Ninety-five Theses* to the church door; Vasco da Gama has not yet sailed around Africa into the Indian Ocean; and Christopher Columbus is a toddler. The art and science of medicine are still familiar to our unwell voyager. The fifteenth-century physician is still the product of a university, still a scholastic, still a Galenist, still drawing horoscopes to assist with diagnosis, and still designing treatment to rebalance a patient's humours. Bloodletting is still a standard procedure and medication is still made from materia medica.

The market for medical care is changing, however: male physicians are beginning to face real competition from male apothecaries. By the 1450s, although they continue to sell other items like soap, spoons, and wine, apothecaries are involved in making and selling medication. Frequently, they also offer medical advice. In London, apothecaries annually appoint two inspectors to make sure that every one of them is selling good-quality ingredients and compounds. Although they operate within the guild system, which means training apprentices to become journeymen (men no longer needing training but not yet proficient enough to be masters) and masters (men who have mastered the craft), paying dues and so forth, apothecaries are still not their own guild. Some of the guilds reorganized at the end of the fourteenth century; the pepperers, spicers, and apothecaries joined the new Company of Grocers ("grossers") for people who deal in bulk quantities ("in gross"). The Company of Grocers is not a good fit for apothecaries, a fact that is becoming more apparent over time.

The third type of practitioner (and the longest-standing one), the housewife, continues quietly and steadily to provide the best, most reliable treatment in England. If she has a garden, she is still growing medicinal plants like agrimony, columbine, elecampane, and hyssop. Everything grown at home has some value to health. As John Smith enthused in 1670, parsnips are "very useful in a Family, being good and wholsome Nourishment, and fatneth the Body much; the Seed hereof being drunk, cleanseth the Belly from tough Flegmatick matter therein". What cannot be grown at home is collected in the wild. Women teach each other where and when to find these ingredients: selfheal in meadows, sorrel in fields, butterbur on riverbanks, St John's wort from hedges and bushes, foxglove in country lanes, betony in the woods, moonwort in rocky pastures, and wild parsley on dunghills. If it grows within walking distance, a woman knows whether to use or avoid it.

Another journey through time brings the sick traveller to the middle of the sixteenth century. Mary I, daughter of Henry VIII, is on the throne. London has grown so far beyond its walls along the Thames that it has merged with Westminster, home to Her Majesty and the eponymous Abbey. The south bank of London – Southwark – has an arena devoted to bull-baiting and another to bear-baiting. In less than a decade, Mary I will be dead and the last of Henry VIII's children, Elizabeth I will ascend to the throne, but it will be another few decades before William Shakespeare takes the stage and has his first play performed. In the meantime, a revolution has just begun on the continent. According to the familiar so-called "Tale of the Triumph of Modern Medicine", before the Scientific Revolution – those centuries our ailing English aristocrat has just moved through – nobody knew anything about medicine and people had to resort to the foul concoctions of equally foul old women. Then modern science came along and revealed real cures. The foul concoctions were tossed away in favour of anti-biotics and the foul old women were charged with witchcraft and burned to death. All hail, the Scientific Revolution and the brilliant, brave men who made it happen.

Versions of this fairy tale have persisted for a long time, and it is time to replace them. Unquestionably, the Scientific Revolution was a period and a movement of tremendous change and many staggering achievements. No aspect of life went untouched: education, the body, politics, religion, gender, race, architecture, technology, the cosmos, and of course illness and health. Science's revolutionaries were also heavily indebted to that vital new technology, the printing press. It enabled the (comparatively) speedy mass production of anything involving words. Ideas could move from one end of Europe to the other with stunning rapidity for the times, and everyone could be sure of having exactly the same text. If a book or treatise could not be had in one place, like London or Madrid, it could be sent from Paris, Leiden, or Milan. Printed matter reached a much larger audience in far less time than any handwritten book, it could be available in many places all at once, and it was less expensive than a manuscript. From the very beginning, it was the most valuable and effective weapon that science's revolutionaries had. It arrived none too soon in England, around 1476, courtesy of William Caxton.

From medicine's point of view, the Scientific Revolution begins in the sixteenth century when the failures of Galenic medicine become too much for two very different men: Philippus Aureolus Theophrastus Bombastus von Hohenheim (c. 1493–1541), who took the name Paracelsus later in life, and Andreas Vesalius (1514–64). They have different frustrations and different solutions, but they both aim for comprehensive change. Born in Switzerland, Paracelsus is educated at a charity school where boys are taught mining, geology, and metallurgy. He attends universities throughout central Europe, earns a medical degree and, convinced that experience is a better teacher than books, starts travelling to learn as much as he can. By the time he returns to the Alps, he has formulated a revolutionary set of ideas. Paracelsus replaces scholasticism with observation and experimentation, rejects the whole humoural theory, maintains that disease infects a body from without, and promotes chemical medicine. He believes that diseases disable organs, which then produce poisons that cause illness. To cure the body's self-poisoning, the body must be dosed with poison (the theory that "like cures like"). As eighteenth-century physician John Fothergill understatedly wrote in his lecture notes: "The vegetables were more used by the antients [sic] than the Paracelsists." Galenists think that this approach is madness, if not murder, but Paracelsus argues that his medications are perfectly safe as long as the dosage is properly calculated.

This is Paracelsus's most significant impact on the history of medicine: the elevation of chemical medicine. He did not invent it, but his advocacy set chemical medication on the road to seeming superior to herbal medication. Strictly speaking, there is no such thing as chemistry in the fifteenth century, but there is alchemy, and Paracelsus adapts it to his purposes. Alchemists hold that matter consists of three building blocks: salt, sulphur, and mercury. Their interest lies in breaking down a compound into its building blocks to see how it was made, and then recombining those building blocks to make a better substance. Paracelsus advocates metals as medication because metals are excellent poisons with which to treat the body's self-poisoning. He adopts alchemy's techniques for working with metals (also sulphur and mercury) to create those medications. If that case of syphilis just will not go away, a little mercury should do the trick. All very logical for a revolutionary who, in the words of Allen Dubus, was "influenced by traditional alchemy, by medical theory and practice, and by Central European mining techniques".

Andreas Vesalius is also frustrated with Galenic medicine and scholasticism, in his case because of the frequent disagreement between what the authorities like Galen said about the human body and what he has observed for himself. He is not the first to confront this issue, but in the past, thinkers had come up with complicated solutions to reconcile authority and experience. Vesalius's solution is to reject a millennium and a half of tradition and learning. He concludes that the revered ancients were wrong about human anatomy and devotes years to dissecting human bodies and drawing everything he sees in meticulous, beautiful detail. In 1543 he publishes *De humani corporis fabrica libri septem*, a literally and figuratively tremendous book that reveals what the human body actually looks like from the surface of the skin down to the network of veins and arteries. It still dazzles, five centuries later. Equally stunning are the acts that produce it: Vesalius rejects the ancient writers who had been relied upon for centuries and trusts only what he can see and prove for himself. With *De Fabrica*, Vesalius lays out a new method for identifying knowledge: observation, investigation, corroboration, repetition, and dissemination. Begin by questioning everyone and everything, perform observations or experiments and record the data, use the results to forge an explanation of the previously inexplicable phenomenon, get others to confirm the data, and then share what was done and learned so everyone else can learn what is true.

In England, our ailing time traveller can observe the influence of these ideas by speaking with Sir Francis Bacon (1561–1626), first at the court of Elizabeth I and then at the court of James I. Bacon is inspired by the new method for identifying fact and establishing knowledge. He imagines an ideal society in which institutions exist solely for the purpose of making observations and doing experiments, and asserts that knowledge should be used to benefit humanity. His main interest in the New Science is the relationship between reason and perception. As he points out, the mind can affect the information gathered by our senses – human sensory perception is limited and those senses can trick the mind. Bacon offers several solutions to these problems. Beliefs and expectations, he insists, "must be abjured and renounced with firm and solemn resolution, and the understanding must be completely freed and cleared of them". He also holds that if at least one other reliable, unbiased person saw it, got the same results from doing the experiment exactly the same way, or calculated the same answer using the data, then the finding is fact rather than belief (or hope, or aspiration). In the middle of the next century, Robert Boyle will call this objective, observant person a "modest witness". Two hundred years later, the name for such a person will be "scientist".

By the end of the sixteenth century, proponents of the New Science are on an inventing spree. In addition to books, there are now microscopes (from 1590) and telescopes (from 1608). Technology supports experimentation, which allows humans to see how nature works and what it is made of. For physicians of the Scientific Revolution, experimentation first and foremost means dissection, which is a huge boon for obvious reasons. Between 1618 and 1628, William Harvey (1578–1657) dissects scores of creatures to discover that blood circulates through the body due to the contraction of the heart muscle. When he publishes *Exercitatio anatomica de motu cordis et sanguinis in animalibus* (*On the Motion of the Heart and Blood in Animals*) in 1628, it is a watershed event on the level of Vesalius's *De Fabrica*. Like Vesalius, Harvey reveals how the body actually works and like Vesalius, he exposes impossibilities in Galen's explanations. Also like Vesalius (and many other revolutionaries, scientific or otherwise), Harvey does not receive a wholly warm response to his work. Vesalius's mentor calls him a madman, while one eminent anatomist calls Harvey an "accountant".

On the face of it, the Scientific Revolution seems like the breakthrough that our English patient has been seeking. A whole school of physicians and apothecaries is forming around the New Science. They doubt or have rejected Galenic medicine, they are excited about the new emerging field of chemistry; they are dissecting, recording, publishing, and disseminating; they are writing and visiting like-minded colleagues throughout

Europe; and they are rethinking surgical procedures and medical school curricula.

The entire medical profession is also changing. In 1518, Thomas Linacre, Henry VIII's personal physician, persuades the king to grant physicians a royal charter, creating the RCP. His Majesty follows the charter with an Act of Parliament in 1523, putting the RCP in charge of medicine throughout the kingdom. "Medicines cannot be rightly used, but by them that understand the whole methode of Physicke," as Eleazar Duncan puts it. By the middle of the century, the College's physicians have established rules for licensing and standards of medical practice, and they are prosecuting those who do not meet them. From the beginning, although with irregular intensity over the years, the RCP prosecutes women within London and up to seven miles from the city who have established themselves as female physicians or practitioners of physic. Women, after all, are trained from girlhood to treat illness and make medication, their reputation is superior to that of the professionals, and there are many more knowledgeable women even in a city like London than there are university-educated physicians. Generally, the RCP is more willing to look the other way when it comes to women treating the poor at the request of parish officials, who are legally responsible for caring for the indigent; however, where a woman is a threat to reputation or income, she is chastised and forced to stop.

The apothecaries undergo a change very similar to the physicians. In 1617, Queen Anne's apothecary, Gideon de Laune, persuades the king to give the apothecaries a royal charter and make them a guild. The Company of Grocers, to which the apothecaries belonged, is furious; it means that some of their most affluent members are breaking off, with a consequent loss of revenue, power, and prestige. The RCP is also furious. The apothecaries are serious financial competition. Already they have been cutting into physicians' incomes by making and selling medication, touting their own expertise, and even treating patients. The charter given by James I does not help. He puts apothecaries in charge of medication, although physicians have already been in charge of medicine throughout the kingdom for the last century. So, to which group does medication belong? It is more than a question of territory and prestige. In addition to fees, by this point medication is a significant source of income for physicians and the primary source of income for apothecaries. The fight is about money and authority.

In addition to anti-woman prosecutions, things are changing for women during the sixteenth and seventeenth centuries. Men like Vesalius and Harvey exploited the advantages offered by the printing press, but the new technology is also a powerful tool for women in Europe. In Italy, Isabella Cortese (fl. 1561) publishes *I Secreti de la Signora Isabella Cortese* (*The Secrets of Mrs Isabella Cortese*) in 1561, the same year that Francis Bacon is born. Her book is full of recipes for compounding medication, making perfume, and performing alchemy. In France, the famous midwife Louise Bourgeois Boursier (1563–1636) publishes *Observations Diverses* (*Diverse Observations [of Medical Cases]*) in 1609. Bourgeois's expanded edition is published in 1617, the same year that the apothecaries become the Worshipful Society of Apothecaries. On the other hand, it is not possible to find a printed book by a woman about medicine or the New Science in England during the sixteenth and early seventeenth centuries. In fact, it is extremely difficult to find a printed book by a woman at all, and still harder to find one where authorship was confirmed. That is not to say that print culture in England ignores women; far from it. Male authors are writing books for women, some of them explaining how to do domestic things (mansplaining in print began early) and some of them praising women's authority. As Robert Green writes in 1592: "For my self if I be ill at ease I take kitchen physic, I make my wife my Doctor, and my garden my Apothecaries shop."

Despite the best efforts of Paracelsus and his followers, materia medica remains predominantly herbal, and gardening is still a vital domestic skill. Gervase Markham's *The English Housewife* (1615) has a section on raising medicinal plants, starting with how to care for and plant seeds, and another extensive section of recipes for medications. William Lawson's *The Country Housewife's Garden* (1618) is the first book dedicated solely to women's gardening. It is reprinted nearly twenty more times by 1695, making it a very popular, very common book throughout the century. Lawson recommends that women keep a separate garden for flowering herbs, which had primarily medicinal uses, and explains how to care for them.

Women are building their own, vibrant manuscript culture, with acknowledgement and respect for women's domestic medicinal authority at its root. After all, if it is not considered valuable, why go to the trouble and expense of writing it down? Late in Elizabeth I's reign, women start recording knowledge by keeping recipe books: blank books in which they write down their recipes for anything made at home, including food, drink, soap, face wash, ink, preserves, and medication. Every time the woman of the house has a recipe that she wants to remember, she writes it in the book. Women were already sharing recipes orally; now they are sharing them textually. A woman might write her own recipe in someone else's book, or vice versa, often at the owner's invitation. They also pass down these books to the next generation. Some women go on to inherit recipe books over a century old, and they still add their own recipes. Very quickly, the recipe book becomes an invaluable tool for women: a means of taking care of the home, a way to form and preserve community, and a deeply personal and deeply utilitarian possession.

One more particularly important development is taking place: European imperialism. While our ailing time traveller is moving through time and space between 1550 and 1650, the Portuguese, Spanish, English, and Dutch are slaughtering each other and the Indigenous peoples of the Spice Islands for control of the obscenely lucrative spice trade; establishing trading outposts along the coasts of Africa; colonizing the Americas; and killing and enslaving what will eventually be millions of people, turning the remaining inhabitants into captive markets for European goods. Spices once limited to the very wealthy (monarchs sent nutmeg as a wedding gift to each other) are creeping within the financial reach of less affluent, less power-ful people. That means that more women are making recipes requiring spices. Knowledge of medications does not change, but the availability of expensive ingredients does.

All this is very promising, but the formerly medieval, long-suffering, much-travelled aristocrat we have been following has one last hurdle to clear before the Scientific Revolution arrives in Britain: civil war, which breaks out in the early 1640s. War is, to put it mildly, a significant obstacle to the development of intellectual thought and international trade, the national economy, the arts, religious toleration, and other key elements of a society. Combatants in the civil war are the Royalists, who believe that Charles I should rule and has God's mandate to do so, and the Parliamentarians, who believe that government should be controlled by elected representatives and that mon-archs are subject to the law. Many Parliamentarians are Puritans, people who had been considered so crazy and subversive just twenty years earlier that they had to flee persecution to the howling wilderness of North America, where they established Massachusetts and began wiping out the Indigenous peoples. The Parliamentarians secure victory in 1649 when they execute Charles I, thus asserting definitively that man made (and unmade) kings. The fighting and appalling brutality continue briefly in Scotland and a little longer in Ireland, but England establishes a commonwealth in 1649 and a rough stability ensues. That is all the stability that the Scientific Revolution needs.

Much has changed since our patient took ill in the thirteenth century – at least, much has changed for the male professionals of healthcare. Physician and apothecary have become defined professions with guild status, jockeying for control of diagnosing illness, prescribing medication, and making and selling medications. Despite all their knowledge and training, they are fighting over a small pie; despite illness and injury being a constant of daily life, few people can or do pay physicians or apothecaries for care.

By comparison, little has changed for women. Even after four centuries, they remain the trusted source of most medical treatment in the British Isles. They are respected and reliable; their domestic medicine is always immediately on hand and free of charge; and their medications are time-tested and overall, equally or more effective and less lethal than those of their male, professional counterparts. Their domestic medicine is also available to anyone: if you are sick, you get treatment. In modern parlance, medication is a right. Women share their knowledge with one another and between generations, forming and maintaining personal connections and whole communities through the circulation of that knowledge and sharing of skills.

Women are the reason that physicians and apothecaries are not making more money or winning more respect.

But it is 1650. The Scientific Revolution offers new methods, new ideas, new technologies, and new opportunities. Everything is about to change.

2

The Countess of Kent's Recipe Book

6accdae13eff7i3l9n4o4qrr4s8t12ux. This is how Isaac Newton explained calculus to his friend Henry Oldenburg. If it seems confusing, take heart: Oldenburg did not understand it either. The numbers and letters are an encoded anagram of a Latin phrase, *Data aequatione quotcunque fluentes quantitates involvente, fluxiones invenire; et vice versa*, which translates to "Given an equation involving any number of fluent quantities to find the fluxions, and vice versa." Newton had not worked out all the kinks in his mathematical system and did not want someone to use his letter to Oldenburg to figure out calculus first, so he sent his explanation as... this. The eventual fight between Isaac Newton and Gottfried Wilhelm Leibniz over the invention of calculus was ugly. Partisans across Europe, including some of the most important and famous contributors to the Scientific Revolution, took sides while Newton and Leibniz fiercely and impolitely battled for credit. The argument lasted for centuries.

What does it matter who invented calculus? Isn't the point that it was invented? Those are not disingenuous questions;

those are questions about a fundamental principle underlying medicine and medicines: that knowledge can be private property. It can be owned, bought, sold, traded, or made freely available. Therefore, it matters a great deal who invented calculus because calculus and the credit for it not only belong to someone but can also be used as capital. Knowledge as private property was a crucial development in the Scientific Revolution and the engine that fired it. This concept was also an essential tool for male professionals' efforts to take the making and provision of medication from women, and to move it out of the home and away from the community – a story that starts in 1653 with Elizabeth Grey, Countess of Kent.

She would have been no ordinary woman of any age, but she was not quite as unusual for the seventeenth century as twenty-first century readers might expect. She was born Lady Elizabeth Talbot, daughter of Gilbert Talbot, Earl of Shrewsbury, and raised in the court of Elizabeth I. The Tudors believed in education for their daughters as well as their sons, and Tudor monarchs' courts were full of dazzlingly talented and skilled minds. Young Elizabeth Talbot was one of those gifted and fortunate women who received a spectacular education and became fluent in several languages. She was fascinated with new ideas and inventions, and became a patron of the arts at a young age. She was extremely intelligent, highly educated, fabulously wealthy, and a favourite of the queen – not a bad way to start her twenties.

In 1601, Elizabeth Talbot married Henry Grey, who became the 8th Earl of Kent in 1623, making her the Countess of Kent. Henry's most significant accomplishment appears to have been marrying Elizabeth. He hired a good manager for his estates, John Selden, and like many other noblemen sponsored a candidate for the House of Commons – again, John

Selden, who proved a very good choice. Henry Grey served one term in Parliament and two terms as Lord Lieutenant of Bedfordshire, but he always held that office jointly with another peer. He withdrew from political life after 1626 when he was forty-three and did little with his time. He never served the crown in a military, administrative, or diplomatic post; he had no particular talent like training horses or swordsmanship; he was not notably educated; he did not collect anything like art or ancient manuscripts; he did not write or compose anything; he did not make improvements to Wrest Park, the family seat; and he did not travel except from one house to the next. Henry did not impose much on his wife, either, from the very beginning. The Greys did not consummate their marriage for nearly a year after the ceremony, although they were living together, because Elizabeth did not want to. From her activities for the rest of their lives together, Henry made no objection to his wife going wherever she wished and doing whatever she wished.

Doing whatever she wished included carrying on an affair with the aforementioned John Selden, who was as unlike Henry Grey as possible. Selden probably met the earl and countess because he lived almost across the street from their London house in Whitefriars. His portrait at the Yale Law School shows a narrow-faced man with a large nose and sensitive eyes. He was not particularly handsome, but he was brilliant. As a politician, he was often consulted by members of the government to help develop policy. His work on law and the principles of government was required reading long after his death. He knew five modern languages (including French, Italian, German, and Spanish) and was familiar with nine ancient ones (including Old English, Hebrew, Chaldean, and Ethiopic), which must have appealed to the linguist in Elizabeth Grey. His love of literature must also have appealed; he personally knew John Suckling, Michael Drayton, Ben Jonson, and John Donne, and worked as a research assistant for Francis Bacon.

The earl and countess socialized with the most intellectual and intellectually adventurous people of the age. When James I succeeded Elizabeth I in 1603, the Countess of Kent joined his wife's court. In her portrait at the Tate Gallery, painted shortly after Queen Anne's death in 1619, Elizabeth is wearing mourning (see plate section, p.1). (It was Anne's apothecary, Gideon de Laune, who persuaded James I to make the apothecaries a guild.) The countess was also part of the inner circle of the next queen, Henrietta Maria of France, which included Théodore de Mayerne, personal physician to the queen; John Evelyn, noted botanist and Fellow of the Royal Society; John Pell, renowned mathematician and Fellow of the Royal Society; and Sir Kenelm and Lady Digby (about whom you will hear later on). The Greys shared many of their friends with John Selden and with Elizabeth's sister and brother-in-law, Alethea and Thomas Howard, Earl and Countess of Arundel and Lennox, including the Digbys, Ben Jonson (also a very good friend of Sir Kenelm Digby), and Sir Robert Cotton. Sir Robert learned Italian from John Florio, who had taught Elizabeth Grey, and owned the only known manuscript of *Beowulf*, which was almost destroyed when his library caught fire in 1701 (visitors to the British Library can see the charred edges of the pages). London was much smaller then, and the group of people with the same level of education who shared interests in learning was very small indeed.

When her husband died in 1639, the countess became even more independent and wealthy. She lived openly with John Selden and continued to socialize with intelligent, interesting people at the leading edge of cultural life. She conducted her own investigations into the workings of nature, a popular activity for educated male and female aristocrats at the time. (Her sister Alethea, who worked with Elias Ashmole, another member of the Scientific Revolution and founder of the Ashmolean Museum in Oxford, built and equipped her house, Tart Hall, so she could do experiments.) Elizabeth Grey acquired a reputation as a paragon of elite femininity: generous to friends and needy artists, uninterested in court intrigue, and occupied with the new, revolutionary ideas, although only at home. In contrast, Margaret Cavendish, Duchess of Newcastle, was also occupied with the new, revolutionary ideas but insisted on barging into male-only spaces and publicly rebutting men's ideas.

Notwithstanding her staggering wealth and scads of servants, Elizabeth Grey was the woman of the house – or in her case, several houses. For centuries, "woman of the house" was an executive position. Regardless of the resources and size of the family, from the lowliest labourer to the wife of an earl, the woman was expected to manage the household and look after her family's well-being. "Family" was not just blood relatives; in the seventeenth and eighteenth centuries, the term included servants, employees like estate stewards, apprentices, journeymen who lived with the master's family, and even tenants. A housewife was expected to keep the household accounts; the wives of tradesmen often also kept the business's books.

For the vast majority of women who did not have servants, daily life was an endless train of physical labour. Domestic tasks included cleaning (house, furniture, household goods, people, clothes, linen, and so forth), making clothing (spinning, weaving, sewing, knitting, and embroidering), repairing things or arranging for their repair, tending the kitchen garden, stocking the house with food (harvesting, drying, preserving, pickling, curing, candying, brewing, fermenting, and buying what was not grown or raised at home), cooking, bearing and raising children, caring for her spouse, and in agricultural families, caring for livestock and helping with the harvest. A widow would also be responsible for helping her children find suitable spouses, which in affluent, especially noble families could mean selecting the person for them. Regardless of marital status, every woman was also responsible for medication. As *The Ladies Dictionary* (1694) explained, the virtuous housewife's "Closet is stored with *Physicks* and *Cordials* prepared with her own *Skill* and *Industry*, to send to her poor Neighbours when they are sick or in pain". Gathering, storing, sometimes buying, and using ingredients was just the first stage; preparing medications "with her own *Skill* and *Industry*" was the next, and administering it the last.

The higher up the social ladder she was, the fewer of these tasks a woman did herself unless she chose to. Katherine of Aragon, first wife of Henry VIII, daughter of Queen Isabella and King Ferdinand, and one of the most educated women in Europe, made and embroidered Henry's shirts herself. A woman like the Countess of Kent would not peel potatoes or poach fish, but she would collaborate with the cook to design menus and develop dishes, supervise the housekeeper to ensure that the house was running smoothly, and check that the servants were doing their jobs and being treated properly. Household expenses were her responsibility, and those expenses could be complicated and extensive at each house. When the Duke of Beaufort undertook a six-week progress through Wales, the Duchess of Beaufort kept track of the expenses for a retinue of more than fifty people. Hosting an important visitor like the Duke of Beaufort required weeks of preparations by the female head of the house. Even changing residences could involve fearsome logistics.

Housekeepers were at the top of the servants' hierarchy and were expected to know everything about maintaining a house that a poor woman or tradesman's wife had to know. In *The Ladies Dictionary* (1694), the job description listed "Competent Knowledge in Preserving, Conserving, and Candying, making of Cates [cakes], and all manner of Spoonmeats, Jellies, and the like: also in Distilling all manner of Waters", as well as "competent knowledge in Physick and Chirurgery, that they may be able to help their maimed, sick, and indigent Neighbours; for commonly all good and charitable Ladies make this a part of their Housekeepers business". A noblewoman like the Countess of Kent would keep track of the medication supply in any of her houses as she would keep track of anything else, like the silver, the linens, or the amount of wine in the cellars.

To manage all of the information about what was or could be made in the kitchen, women kept recipe books. Previously, recipes were part of an oral tradition. Sometimes they took the form of rhyming verse because it was easy to remember. Jane Barker, one of the two central players of Chapter 4, includes rhyming recipes in *A Patch-Work Screen for the Ladies* (1723) as evidence of her protagonist's domestic excellence. "The Receipt for Welsh Flummery [a sweet, jelly-like dessert] / Made at the Castle of Montgomery," for example, sounds almost exactly like a recipe in prose:

Take Jelly of Harts-horn, with Eggs clarify'd,
Two good Pints at least; of Cream, one beside.
Fine Sugar and Limons [*sic*], as much as is fit
To suit with your Palate, that you may like it.

Three Ounces of Almonds, with Orange Flow'r-Water,
Well beaten: Then mix 'em all up in a Platter
Of China or Silver; for that makes no matter.

Written recipes and recipe books became popular around the
end of the sixteenth century among women who were literate
and could afford a large blank book, its binding if one were
needed, pens, ink, and time. Even these requirements could be
worked around: the first part of Mistress Honore Henslow's
"book of surgerie and phisick" from 1601 was copied down
by her servant, Andrewe Plowden, although the reason is not
given. This period for recipe books roughly coincides with the
second half of Elizabeth I's reign, just around the time that
Elizabeth Grey, Countess of Kent reached adulthood. Recipe
books contained the instructions for making household
goods at home. Items like candles and ink, cosmetics like face
washes and rouge, and foodstuffs like roast chicken, braised
turnips, currant wine, pickled plums, and rum cake were
all confected in the kitchen, so their recipes would be in a
recipe book.

Medications were among these substances: cough drops,
burn salve, elixir to reduce cramps, drinks to cure jaundice,
poultices for cuts and blisters, and plasters for fevers were
also made in the kitchen. Under C in the table of contents,
one seventeenth-century book lists, in this order, one recipe
for "chocolate puffs", one "for ye Collick", two recipes "for a
Cough", and one each "for a consumption", "a good Cordiall",
"a good cake", something called "spewsbry cakes", "for a
cake", and "consumption". An early eighteenth-century recipe
book (c. 1725) listed together recipes for baked goods, preserving fruit, and medication, including a treatment for strangury
(frequent, painful urination in minute quantities at a time).

Anything having directly to do with the management of health – whether to restore or to maintain it – was made from a recipe; even an aristocrat like Elizabeth Grey, Countess of Kent, would need a book to keep track of the good ones.

As utilitarian, household objects, recipe books did not leave the home any more than mattresses and chamber pots did, but they were also deeply personal items and cherished family heirlooms. They had a far larger significance to individuals, to families, and to the larger society than just as collections of instructions for manipulating vegetables and meat. A woman treasured her recipe book. It was a valuable compendium of information. Tired of cooking carrots the same way? The recipe book would have several ways to prepare them. Someone in the family has an upset stomach? There would be at least a few recipes for remedies. Need an impressive dessert to serve guests? Or heartburn medication when that fancy dinner is over? Check the recipe book. The supply of wines is running low? A parent is going blind? The baby has worms? You can't get pregnant? There was a recipe for every concern, and if a woman lacked a treatment for kidney stones or a good way to cook venison, a female relative, friend, or neighbour would have one to share.

Recipe books were organic creations, always added to and amended. Physically, they were books of blank pages, many fairly large, and recipes were inscribed wherever the owner wanted them. That practice produced highly individualized recipe book organization, to put it mildly. As earlier examples suggest, some women just added recipes one after the other, so instructions for making pea soup could follow instructions for an earache treatment. In 1692, Mary Harrison wrote instructions "To make Barly Broth" right after those for "A Diet Drinke for the Wind". One eighteenth-century writer entered "To Roast

Hamm", "To make Ink", and "To make Black Salve" one after the next. Other recipe books were more ordered. Some women grouped recipes by kind: recipes for baking in one section and for preserving in another, or recipes for salves in one place, for cordials in another. Some women made a table of contents in the front, some kept one in the back, some had none at all. Sometimes recipes were listed in alphabetical order, although that did not inevitably offer much organization. Was that salve for gunpowder burn listed under S for salve, B for burn, or G for gunpowder? The answer depended on the woman who listed it.

Women were always in search of better recipes, whether it was a better way of pickling plums, a tastier fricassee of rabbit, or a more effective medication. They were empirics: people who used trial and error to reach conclusions. Burn medicines are common in recipe books, unsurprising for a population that spent a lot of time in the kitchen. Recurring recipes reveal family health over the years. A cluster of recipes for eye problems in one Mrs Johnson's recipe book reveals that she tried for more than a decade to find a successful treatment for her family's recurring eye problems. Empirics were not the same as empiricists, which early scientists took pains to establish. Empiricists observed or experimented on nature to understand how it works. Empirics were also not the same as theorists, the physicians who held a theory about the body's workings (like, say, Galenic medicine) and forged treatments in accordance with the theory and at times in discordance with observed reality.

The empiric method worked pretty well overall and far better than theoretical medicine like Galenism. It is not fair to judge anyone's success rates against major illnesses that have only just been (or have not yet been) conquered like smallpox, plague, or cancer. It is much more reasonable to judge success rates for other problems like heart palpitations, fever, irregular menses, jaundice, and fluid retention. If a member of the household told their wife, mother, aunt, or sister that they had a headache or a cut, there was a good possibility that they would get a remedy that actually remedied. At the time, Jesuit's bark or cinchona was used to treat fevers; in the twentieth century, chemists discovered that it contained quinine, which is good for treating malaria. The bark of the willow was used to reduce fever; it became the foundation for aspirin.

Useful recipes and up-to-date information tailored or tailorable to the needs of the family were all good reasons to value a recipe book, but the books offered far more than that. They created a community of knowers and promoted an ethos of sharing – the heart of domestic medicine. No woman, including the Countess of Kent, kept her recipe book in isolation. Women exchanged recipes all the time, whether casually, such as over the garden fence or during a chat at the market, or more formally, such as by invitation. They copied down recipes for food and medicines from dictation or asked the original source to write it in their books. Visitors and honoured guests could be invited to contribute a favourite recipe.

This practice of sharing and collaborating makes recipe books a place where the ordinary women of history, normally invisible and disregarded, become visible and vital. Over and over again, I have gingerly turned the first page of a recipe book three or four centuries old and fallen into the life of a woman, a family, and a community of friends and neighbours and distant relatives. A veritable flock of Mrs Hodgsons contributed recipes to Mildred Hodgson of Liverpool's book, including Mrs Hodgson from St Albans (about 180 miles distant), Mrs Hodgson from Ashbourne (about 65 miles), and Mrs Hodgson from "Crape Marsh" (possibly near Sheffield and therefore around 80 miles away). Even an unmarried relative, Miss Hodgson from Manchester (about 40 miles), contributed a recipe. Mildred also had visitors from far-off places. Her book includes recipes from Mrs Webb of Hillingdon, Mrs David of Calwich, Mrs Leigh of Greenhill, Mrs Mills of Norbury, and Mrs Pyshe of Noonham. A pair of surviving recipe books from the Wise family reveal that Prudence Wise's daughter-in-law, Mary, copied a long sequence of recipes into her own recipe book directly from the elder Mrs Wise's. The younger woman had already begun collecting recipes, but the addition of her mother-in-law's recipes early on offers a picture of the two women working together. Whether the copied recipes are a sign of harmony is another question.

Not every recipe was attributed or signed. After reading hundreds of sixteenth-century, seventeenth-century, and eighteenth-century English and Scottish recipe books, I found only one with every single recipe attributed to someone. To some extent, including a name depended on the personality of the person keeping the book or on the personality of the one giving the recipe. Contributions from a member of the landed gentry or aristocracy always included name and title, a practice that appears regardless of the century, place, or class of the book's owner. In some books, the contributor's name appears only at the recipe and not in the table of contents. The author who identified the source of every recipe wrote the name next to the binding, to the right or left of the recipe. Elizabeth Smith's recipe book, started in 1700, does not have a table of contents, but she wrote the title of "My Lady Greshame's Almond puddings" in larger letters and with a differently shaped quill than she wrote the directions, which makes the words stand out quite clearly. She did not have very many recipes from titled personages, so it's no wonder she showed it off.

What's startling is when a recipe from a physician or apothecary turns up:

Strengthening Plaister
℞ Rx Empt ad Herniam ℥vi
   Paracels
Diasaponis a ℋℨ ss
ℳ of extend sup AluCa
   Φ tergo

As I will discuss in Chapter 6, the inclusion of professionals' recipes was significant for a number of reasons. For now, the notable issue when professional language appears is how disruptive and intrusive it seems. The same holds for the appearance of someone professionally credentialled in a group of laypersons. No matter how distinguished the source or how many professional sources a book contained, the recipes from friends and neighbours far outnumbered them. Like people "of quality", physicians were always identified as physicians in recipe books. Apothecaries, who did not have the same social cachet because they were guild-trained rather than university-educated, were only sometimes identified as such, but the language of their recipes, like this one, often provides that information or at least strongly suggests the supplier. Even when a male professional's recipe was recorded in standard, written English, it often looked and sounded different because of its technical lexicon, such as the recipe for a powder to treat a toothache:

take [*sic*] red coral powderd 2 drames, Bole one dram, mastic half a dram, scutchinnel half a scruple, of burnd allum half a scruple, dragons blood half a scruple, Acorn cups A scruple, make all this into a fine powder & rub y^r teeth with it twise a week.

Elizabeth Smith's recipe for "The greater Palsie wat^r & also Apoplexye draught by Mr Mathias", another apothecary, was two or three times the length of all the other recipes in her book.

There has been little attention to these recipes from male professionals: how women got them, why women got them, from whom women got them, what a professional's motives were in allowing a woman to have them, or what the appearance of an apothecary's or physician's recipe in a woman's recipe book can tell us. At this stage in the story, I want to make clear two points. The first is that professionals' recipes were not always translated from the official apothecary code into English; the second, that women sometimes paid for the recipe, translated or not. The keeper of one seventeenth-century recipe book spent £100 for a recipe, a staggering sum for the time and roughly equivalent to £25,000 today. Part of what makes the appearance of a professional's recipe so shocking is that it is the equivalent of Isaac Newton's formula in his letter to Henry Oldenburg: the sign of a system in which knowledge is owned, bought, sold, and exchanged, rather than freely given. Prescriptions and professionals' recipes are Trojan horses, seemingly a gift to the woman but in reality an invasion of forces hostile to domestic medicine, women's knowledge, and women's community.

That is not to say that women did not recognize their recipe books as objects with value. A woman's recipe book was the ultimate training manual for future women in the family. The Johnson family began their recipe book at the end of the seventeenth century and only stopped keeping it at the start of the nineteenth. John Gibson began a book of medicinal recipes in 1634; Joan Gibson acquired it and started adding culinary recipes in 1669; and Joanna Gibson continued it, starting in 1708. One woman specified in her will that her recipe book should go to her daughter, suggesting that other relatives might try to make a claim. Another had the foresight to have her recipe book copied during her lifetime so she could bequeath one copy to her daughter and the other to her daughter-in-law. What went into that book was knowledge, and for the women who kept recipe books, knowledge's value lay in its use. Distribution was a vital element of use. Knowledge of useful recipes was meant to be shared with the community.

That point is underscored by the fact that the vast majority of recipes are undated, unsigned, and unattributed, not just in any given book, but throughout the genre. The many different handwritings for individual recipes in recipe books confirm that women often wrote in each other's books and without the need for credit. So even though Mrs Bodenham knew the recipe for "Lemond water" that she gave to Mrs Brooke, Mrs Bodenham did not own it. She never would have charged anything, let alone £100, for any of her recipes, no matter how effective or delicious. The Countess of Arundel and Lennox gave her sister, the Countess of Kent, her recipe for manchet, a very fine kind of bread. She certainly did not expect goods or coin in return. Recipes for food and medicine were circulated without restrictions, and the sharing of this knowledge bound women together across class, age, time, and space. Elizabeth Smith was not a countess like Elizabeth Grey, but they both needed recipes for medication to treat the people in their households.

This was how domestic medicine worked. These connections were far more than networks of knowledge transmission. They were personal and intimate. Every recipe was handwritten. Sitting down with a recipe book is sitting down with every woman who wrote in it. Handwriting styles changed significantly between the early seventeenth century and the mid-eighteenth century, so even an undated book offers evidence of when writers learned penmanship. Two recipes in two different styles from two different centuries next to each other in a recipe book tell a story of two different generations writing in the same book. Handwriting is also highly personal. Letters themselves – the size of a loop in a P or R or D, the height of an L or an I, the number of separate strokes used to make an M or W – introduce a new contributor of knowledge, even if her handwriting is all that can be known about her. Turning the pages of a recipe book is very often travelling through at least one woman's life. Her handwriting changes slowly from a young woman's steady, clear hand to the uncertain letters of arthritic fingers. Names or other hands recur frequently for a while, taper off, and disappear, as different women are more or less important or present in the recipe keeper's life.

For the owners of recipe books, leafing through the pages could remind them of mothers and grandmothers, sisters, cousins, aunts, and friends – women perhaps no longer there to share the kitchen with. A particular recipe might call to mind an event or a place, an occasion when women were gathered together to taste or write or tinker with a recipe. Those letters and words were placed on the paper by hands that stirred pots, turned meat, bathed babies, and wiped tears. Innumerable women left no trace at all that they ever existed, but recipe books provide evidence of thousands of women who formed communities and shared their knowledge to care for others and help one another.

Domestic medicine situated illness and treatment within communities, from households to neighbourhoods, from extended family to distant friends. A woman never treated sickness alone, a sufferer never ailed alone, and sickness itself was always personal and never solitary. The ten-year-old apprentice could expect the same medical care from the wife of his master as the forty-year-old master himself. The lady of an estate was expected to treat her impoverished tenant as well as her own new baby and her maid. Sometimes, a particularly talented woman would become the de facto "mediciner" for her community, the same way a woman could find herself the local midwife. She might charge people for medications in that case, but she was extremely unlikely to turn anyone away. If you did not have a penny, you might have eggs or be willing to chop wood. Recipe books were the fulcrum of domestic medicine's concept of illness, embodying sharing, knowledge, family, expertise, and connection.

That is why when *A Choice Manual of Rare and Select Secrets in Physick and Chyrurgery Collected and Practised by the Right Honourable, the Countesse of Kent, Late Deceased* was printed in 1653, no one was surprised that one of the richest women in England, the Countess of Kent, had a recipe book. A woman had responsibilities, even a countess. Everyone wanted to see what such a celebrity knew and did. *A Choice Manual* flew off the shelves and it was well worth the price. It was packed with cures for every condition from bad breath to vomiting. There were recipes for medicines to treat dog bites ("Take Ragwort, chop it, and boil it with unwashed Butter to an Ointment"), kidney stones (the herbs only work if picked in June or July), and tuberculosis (start with a live rooster).

It also had recipes for medicines to treat specifically female complaints. "For a dead Child in a Womans Body", the countess advised to "Take the juice of Hysop, temper it in warm water, and give it the Woman to drink" to expel the foetus. The next recipe, to ease frequent, heavy, or irregular menstruation, instructed readers to "Take a Hares foot, and burn it, make pouder of it and let her drink it with stale Ale." There were recipes for medicine to induce menstruation, to help a woman expel afterbirth, and to prevent miscarriage while travelling, a reasonable concern in an era before shock absorbers and road repair. If some of these recipes were used to induce miscarriage, they were labelled circumspectly.

Her ladyship's recipes were consistent with what male professionals could have prescribed, although there was so much variation in medications for conditions at the time that it would take a lot for a medication to seem truly bizarre or innovative. *A Choice Manual*'s recipe for heavy menstruation, for example, was essentially the same at Thomas Brugis's recipe in *The Marrow of Physicke* (1648) "To stop womens immoderate fluxe": "A Hares foote, and burne it to Powder, and drinke it first, and last in stale Ale, till you be whole." Like the Countess of Kent, Christopher Wirtzung recommended a drink with hyssop to expel a dead foetus, although he preferred wine over water and noted that vervain and myrrh work equally well. On the other hand, Wirtzung also suggested an "external application": "One may also hold before the privities swines bread, or cotten wooll made wet in the juice of the same: and to put it before into the bodie is also marvelous good for this purpose." Elizabeth Grey had a second recipe for treating "immoderate flux", which in contrast called for powdering date pits, cumin seeds, "Grains", and saffron together and serving a spoonful in warm malmsey, a kind of wine.

Holding "swine's bread" between a pregnant woman's legs is not likely to do very much; Grey's prescription of large doses of saffron, on the other hand, can induce contractions and cause miscarriage.

The second part of *A Choice Manual* provided recipes for edibles and drinkables served at mealtimes. Entitled "Most Exquisite Waies of Preserving, Conserving, Candying, &c.", it told the "true delighted gentlewoman" how to manage foodstuffs. These are not dainty recipes. "To sauce a pig" begins: "Cut off the head." There are elegant dishes, showing the reader how "To boyle Flounders or Pickerels after the French fashion" and "To roast a Pig with a Pudding in the belly", but most are far sturdier: "How to roast a Leg of Mutton", "How to roast a neck of Mutton", "How to stew a Rabbet", "A Potato Pie for Supper", and "To make Cheese-cakes". The book, in other words, really is for running a household. "To make a Pudding in haste" sounds like the kind of recipe a working parent might use and "To make a Leg of Mutton three or four dishes" is the sort of help a frugal housewife might value. *A Choice Manual* also has a large selection of recipes for preserving fruits and vegetables so they can be eaten out of season. "To keepe Quinces all the year" is followed by recipes for pickling (for instance, "Cowcumber", purslane, "Broom-bids", and oysters), preserving (including oranges, green walnuts, white quinces, red quinces, raspberries, gooseberries, mulberries, apricots, unripe white damsons, ripe damsons, white pippins, grapes, currants, and medlars), and candying (apricots, pears, plums, lemons, "Ringo roots", and "all kind of flowers in ways of the Spanish candy"). Both the method of cooking (boiling, stewing, roasting) and the ingredients (pig, pickerel, beef) were every woman's method. There was nothing remarkable about the recipes per se, with a few exceptions.

When Elizabeth Grey's recipe book was published, printed books of recipes for medication, cooking, or both had been around for a long time. There was Bullein's *Bulwarke of Defence Against all Sicknesse, Soarenesse, and Woundes,* which came out in 1562, as well as *Approv'd Medicines and Cordial Receipts* (1580) by Thomas Newton, *Prepositas His Practice* (1588) by "L. M.", *An Alphabetical Book of Physicall Secrets* (1639) by Owen Wood, and *A Pretious Treasury* (1649) by Salvator Winter and Francisco Dickinson. *A Proper New Booke of Cookery* (1575) not only provided culinary recipes but also listed which recipes were appropriate for which course; however, it did not offer any recipes for medication. *The Good Huswifes Handmaide for the Kitchin* (1594) was an impressive collection of cooking recipes in no particular order, while *The Widowe's Treasure* (1588) offered "sundry precious and approoved secrets in Phisicke, and Chirurgery" and "sundry pretie practices and conclusions of Cookerie", all mixed together like any other recipe book. But these books were not written by women, and even anonymous books had no sign of a woman's hand anywhere in the composition. The purchaser of *The Good Huswifes Handmaide for the Kitchin* would assume that it had a male author because only men's work made it into print.

The Countess of Kent's *A Choice Manual* was unlike any of these books. It had all the features of a typical manuscript recipe book, with recipes appearing in the order in which they had been added: "A Medicine for the Stone in the Kidnies" is followed by "To make Horse Raddish drinke", "Sir John Digbies Medicine for Stone in the Kidnies", "An excellent Sirrup against Melancholly", "An excellent Receipt for the Plague", and then "An excellent Cordiall". Treatments for the same ailment are scattered throughout the book. Recipes

"For an Ague" are listed on pages 26, 131, 139, and 147; those "For a Stitch under the Ribs" can be found on pages 10, 59, and 95. *A Choice Manual* has a table of contents in the front, grouping the contents alphabetically: all recipes beginning with A are listed together, all recipes beginning with B are listed together, and so on. Like so many books of its kind, the principles governing the alphabetizing are a bit harder to identify. For instance, burns are one thing: "For burning with Gunpowder" and "For a burning with lightning" appear under B. However, bites are another: "For a biting of a Mad Dog" can be found under D. Like some other women, the Countess of Kent recorded which recipes were given to her by other people. The medication section includes "Dr Willoughby's Water" (under W), "Dr Stevens Water" (under S), and "Mr Ashley's Ointment" (under A). In the cooking section, the only recipe credited to someone else is her sister's bread recipe: "Lady of Arundel's Manchet".

All that distinguished the countess's recipe book from any other recipe book was that it appeared for sale in print, under her name, but this exception is everything. Transferring the Countess of Kent's recipe book into print and putting it on sale was a complete transformation of the recipe book. It made something public out of something private; something commercial out of something domestic; something mass-produced out of something individual; something common out of something unique. Turning a manuscript book into a printed book converted domestic knowledge that had been shared freely in a community of women into knowledge that could be obtained only by people with the money to pay for it – *A Choice Manual*, a symbol of domestic knowledge including medicine, had become a commodity. In one move, the Countess of Kent reversed everything intrinsic to her recipe book or to any recipe book. Impressive for a woman who had been dead for two years.

The person who turned her ladyship's private recipe book into public, saleable goods was not her ladyship at all, but a man named William Jervis. Not much is known about him except that later in life he called himself a physician and a professor of natural philosophy. The greatest mystery about Jervis is how he got his hands on the Countess of Kent's recipe book to begin with. None of the records of her household, her friends, or her extended family can account for him. It has been suggested that Robert May, a famous cook, gave it to him. In the seventeenth century, "chefs" were primarily men. They were called "cooks", but the job and the prestige were those conferred on chefs today and so was the gender bias. The person who ran the kitchen in an aristocrat's household was generally male. Robert May had been trained by illustrious English cooks (his father included) and had learned from a famous French chef in Paris. For fifty years, May cooked in aristocrats' households, including the Countess of Kent's. Maybe, the thinking goes, she left the recipe book to Robert May or he somehow wound up with it because they had worked together on the daily menus.

Maybe. It is hard to believe that the countess would give away her recipe book while she was still using it, and he was in her employ in the 1640s, when she was still hale and hearty. Would she have bequeathed it to Robert May, an unrelated male servant? Admittedly, May was very highly regarded as a cook. He was another acquaintance whom she shared with Sir Kenelm and Lady Digby. It is possible that she might have left her recipe book to Robert May if she had a particular fondness for him. Then again, he had not worked for her for nearly a decade when *A Choice Manual* was published in 1653. One thing is certain: whether William Jervis got the Countess of Kent's recipe book from Robert May or he acquired it by other means, Jervis was free to do with it whatever he thought best. And what he thought best was to turn a handwritten recipe book into a printed book attributed to a woman that he entitled, *A Choice Manual, Or Rare and Select Secrets In Physick and Chyrurgery: Collected, and practised by the Right Honourable, the Countess of Kent, late deceased. As also most Exquisite waies of Preserving, Conserving, Candying, &c.*

Jervis was a clever marketer. Celebrity sold as well then as it does now, and the Countess of Kent's name in the title guaranteed sales. (The title page of Hannah Woolley's recipe book, *The Ladies Directory, in Choice Experiments & Curiosities of Preserving in Jellies, and Candying Both Fruits & Flowers* [1662] stated that she "hath had the Honour to perform such things for the Entertainment of his late Majesty".) The first of anything – in this case, the first woman's recipe book to appear in print – is also alluring. The cleverest part of Jervis's title was the phrase "Rare and Select Secrets". The book claimed to contain things previously known only to the countess, rare knowledge that she kept a secret from everyone else. Once a secret is shared, of course, it is no longer secret. By printing a recipe book, William Jervis was exposing a woman's life, her experience and knowledge.

Furthermore, the knowledge in a woman's recipe book was the opposite of secret. It was meant to be shared and shared with anyone who needed or wanted it. Men were not prohibited from learning recipes, not even by cultural expectations, and there are enough recipes attributed to men to make the phenomenon unremarkable, if uncommon. Occasionally, men even kept recipe books, as was the case with the Gibson recipe book mentioned earlier. Physicians' notebooks offer instances of physicians consulting their wives for remedies when their own recipes or knowledge gave out. One son seized his mother's recipe book immediately after she died, despite knowing that his mother had willed it to his sister, because he recognized its value. It took a lawsuit for his sister to recover it. So when Jervis called the countess's recipes "Rare and Select Secrets", he changed their perceived identity from shared, shareable knowledge into knowledge not at all meant for sharing. In printing it, Jervis acknowledged that this knowledge had value. It was also knowledge taken from Elizabeth Grey by a man; knowledge that had been female and communal now belonged to a man as his private intellectual property.

The success of *A Choice Manual* was instant and catalytic. Two editions appeared in 1653, very unusual for the time. Within a decade, the printer Gertrude Dawson was putting out the fourteenth edition; the twenty-second edition appeared in 1726. *A Choice Manual* also led immediately to other, similar publications. *The Queen's Closet Opened* came out in 1655, attributed to Elizabeth Grey's friend, the Queen Mother Henrietta Maria, who was in exile. A look at the title page of *The Queen's Closet Opened* shows that it was the result of efforts by "W.M.", who was almost certainly male (scholars suspect that it was William Montagu, Her Majesty's steward). Another printed recipe book, *Natura Exenterata,* appeared in 1655, ostensibly authored by Elizabeth Grey's sister, Alethea Howard, Countess of Arundel and Lennox (she of the bread recipe). Alethea Howard also was dead when *Natura Exenterata* was published. The title page of the first edition declared that it was "Collected and preserved by several Persons of Quality", and the list of male and female contributors is extensive.

The title page of Owen Wood's *Choice and Profitable Secrets both Physical and Chirurgical,* printed in 1656, claimed rather grumpily that these "Secrets" were "Formerly Concealed by the deceased Duchess of Lennox, and now published for the use and benefit of such as live farr from Physicians and Chyrurgeons". If they were the duchess's originally, Wood did not mention it when the book first came out in 1639, the year the duchess died. Whether the recipes did or did not originate in a woman's recipe book, in the mid-1650s Wood certainly thought that putting a famous dead woman's name on it would be good for sales. The reverse was also true: people were learning that they would and should have to pay for a book full of recipes to make medicine and food.

To the casual observer, all these changes suggest that in the middle of the seventeenth century, women and written forms associated with them had cultural power. The wave of female-authored recipe books in print suggests that women were positively associated with "physick", that is, medication. It would seem that women's medicinal knowledge was now valued. Yes and no. *A Choice Manual* was capitalizing on the strong, positive cultural association of women and medication, while undermining it at the same time. Everyone knew that these women were terminally absent, whether by death or by exile. When it came to disseminating their knowledge and expertise in print, like publishing Elizabeth Grey's or Alethea Howard's recipe books, men controlled everything, men put their names on the book as well as the women's names, and men gained the profit from their publication. The use of insinuating titles perpetuated the conversion of women's ordinary, open knowledge into something unusual, exotic, and rare. Women's domestic medicine became knowledge that was "Too costly to wear every-day", as Beatrice puts it in Shakespeare's *Much Ado About Nothing*, and not "for working-days".

The surge of female-attributed printed recipe books in the

1650s was not an introduction into the mainstream of female knowledge and female methods for acquiring knowledge. Female knowledge *was* the mainstream, already freely circulating and circulating for free as women shared it with each other and used it to care for the people around them. Instead, printing women's recipe books moved those books and that knowledge into a male-controlled commercial sphere. Women's actual recipe books were organic, always growing and changing as women learned new things. Putting that knowledge into print froze it where it was at the moment of printing. William Jervis's *A Choice Manual* became an early salvo in the war against women for control of pharmaceutical knowledge and distribution. He acknowledged the power and authority traditionally granted women, their recipes, and their recipe books, but only to diminish them by sexualizing women's knowledge, by making it exotic and rare, and by nudging it into the male domain. He also branded the expert knowledge of food and medicine preparation in the book (which had been communal) as something secret that he had the responsibility to unlock – and to charge for.

The first printing of *A Choice Manual* in 1653 marks the moment when women's domestic knowledge began to acquire a different set of cultural associations. Women continued to keep recipe books until the nineteenth century, but over time and on the whole, recipes from professionals and from printed sources like newspapers began to appear more often, as I demonstrate later on. Men, on the other hand, produced the printed domestic recipe books, and the knowledge contained in those books belonged to the men who produced them. In the century between 1550 and 1650, fewer than thirty books with the word "recipe" or its variants in the main title went on sale, all of them authored or compiled by men. In the half century between 1650 and 1700, that number was closer to 100, the vast majority authored or compiled not just by men, but by physicians, apothecaries, and that emerging breed, scientists. In handbooks for housewives, which contained instructions and tips for responsibilities such as managing servants, keeping to a budget, setting the table, and comporting oneself virtuously, the recipe section shrank dramatically after 1650, with the medication portion of that section shrinking even more so.

Elizabeth Grey, Countess of Kent would probably have been furious with William Jervis. She had spent a lifetime asserting her independence and autonomy, and he was making use of her recipe book – her knowledge, her experience, her friendships, her learning – for his own purposes and without her permission. She used her wealth to help others and promote knowledge, skill, and beauty; Jervis used her recipe book to help himself and, while he was at it, to denigrate women's knowledge and skill. She was a patron: she made communities by gathering people together, inviting writers and artists to live at her house, people who she found interesting and exciting, who she thought would enlighten others. Jervis sold her book to thousands of individuals, who stayed individuals. *A Choice Manual* turned knowledge into property to be used to profit its owner.

Changing the association of recipe books from women to men was a vital beginning for the scientific revolutionaries. It still left them with everything that recipe books contained: recipes for beer-battered fritters and sauced young pigs, as well as recipes for treating jaundice and miscarriage. To separate what they wanted, medication, from what they did not, food, the New Learning's proponents were going to have to use another strategy. It would involve Sir Kenelm and Venetia, Lady Digby, piracy, two kitchens, a faked death, and a lot of inebriated snakes.

3

Chicken Soup and Viper Wine

very day, Sir Kenelm Digby gave his wife Venetia, Lady Digby a glass of viper wine. He made it himself. He took a few dozen live, poisonous snakes, shoved them into a cask of wine, stoppered the cask, and let it sit undisturbed for a few months, until the snakes were dead and disintegrating. He might have strained the liquid before serving it, although his friend, Alethea Howard, Countess of Arundel and Lennox, did not recommend it in her recipe:

Take eight Gallons of Sack which is the best Wine, and to that quantity put in thirty, or two and thirty Vipers; but prepare them first in this manner. Put them into bran for some four dayes, which will make them scowre the gravel and eathy [sic] part from them, then stop your Vessel or glasse you put them in very close until six months be past, in which time the flesh of the Vipers and vertue of them will be infused into the wine, although the skins will seem full, after which time you may take them out if you please, and drink of the wine when you please best to drink it.

A number of things about this story might raise questions for the modern reader. Why did Kenelm give his wife Venetia *viper wine*? *Why* did Kenelm give his wife Venetia viper wine? And why did *Kenelm* give *his wife* Venetia viper wine? These are not the same question, and they are all important. Answering them explains how another aspect of the change from domestic to professional medication took place: the separation of food from medicine.

So why did he give her viper wine? Viper wine was not as exotic in the seventeenth century as it is in this one. Physicians had been prescribing and administering it regularly to patients with skin conditions since Galen. Lecturing in 1635 on the treatment of tumours, Alexander Read recommended viper wine for leprosy. In 1675, Philip Bellon went even further, claiming in *The Potable Balsome of Life* that drinks made with vipers were useful for treating not only leprosy but also sexually transmitted infections, tuberculosis, fevers, and scurvy. A few years later, viper wine took on other powers: in *Pharmaceutice rationalis, or, An Exercitation of the Operations of Medicines in Humane Bodies* (1679), the distinguished physician Thomas Willis advocated viper wine for strengthening a man's "animal spirits", a use also endorsed by William Salmon in *The Practice of Curing* (1681). Additionally Salmon prescribed viper wine and occasionally viper powder (what he called "viperine medicaments") for convulsions, tremors, and somewhat paradoxically, "Paralytic Distempers". For good measure, he also agreed with Bellon that it could clear up leprosy. The poet John Donne even alluded to such medication in a sermon at St Paul's. Samuel Hartlib, who corresponded with all the great minds of Europe, recorded numerous recipes for viper wine in English, German, and Latin. The *Pharmacopoeia Londinensis*, the official compilation of medicinal treatments used by physicians and apothecaries, included a recipe for viper wine well into the eighteenth century.

Given its supposed ability to restore blemished skin and "the animal spirits", ordinary people often regarded viper wine as a combination of Botox and Viagra. In 1633, Francis Quarles wrote mockingly of "Viper-wines, to make old age presume / To feele new lust, and youthfull flames agin". John Jones, in his play *Adrasta* (1635), disapprovingly associated viper wine with witches and black magic because of its power to transform the body. Unsurprisingly, his play was never staged. It was even fashionable for a time for women at the Stuart court to take a glass regularly.

So there was nothing particularly exotic about viper wine. *Why* did Kenelm give it to Venetia? Perhaps she was physically worn out. She had five pregnancies in eight years. Their first child, named after his father, was born after she and Kenelm were married secretly in 1625. She went into labour after falling from a horse late in her pregnancy; to preserve the secrecy of her pregnancy and marriage, she delivered the baby at home with only her inexperienced maid to assist. Venetia and Kenelm's second child, John, was born in 1628, the day that Kenelm set sail for a two-year voyage to the Mediterranean. After receiving the news, Kenelm wrote Venetia from his ship that she could announce their marriage and then raised anchor, leaving her with two small children but without the financial, social, and personal support of a husband. After Kenelm returned in 1630, Venetia had three pregnancies in the next three years, two of which ended tragically. Her third child, a son named after his grandfather Everard, died within hours of his birth, and she miscarried twins in the seventh month. Small wonder if after so much physical and emotional battering Venetia's "animal spirits" were low.

Or perhaps Kenelm gave Venetia a regular glass of viper

wine to restore or preserve her remarkable beauty. Venetia Digby had been considered one of the most beautiful women in England since her teens. John Aubrey, a seventeenth-century diarist and biographer, called her "that celebrated Beautie". According to Aubrey:

> She had a most lovely and sweet-turned face, delicate dark brown hair. She had a perfect healthy constitution; strong; good skin; well proportioned; much inclining to a wanton (near altogether). Her face, a short oval; dark brown eyebrow, about which much sweetness, as also in the opening of her eyelids. The colour of her cheeks was just that of the damask rose, which is neither too hot or too pale.

Some of Sir Kenelm's biographers assume that he valued her stunning looks so highly that he medicated her to protect them. John Fulton claimed that: "To preserve her beauty, he threw his dynamic energies into experimenting upon her; inventing new cosmetics of dubious ingredients, nourishing her failing strength upon snail-soup and capons fed on the flesh of vipers."* Or his biographers imply that Venetia might have valued her looks so highly that she asked for something to do the job. Both scenarios are certainly plausible, but the evidence does not support either one. Actually, the evidence reveals that biographers and historians have gotten quite a lot wrong about their relationship.

---

* Feeding chopped-up vipers to chickens was supposed to make the latter even healthier to eat. Samuel Hartlib recorded that the Duke of Bavaria's personal physician prescribed viper-fed chickens for the duke, and His Grace was reputed "very vigorous in his health".

For one thing, there is no evidence that either of them ever valued her looks over her character. A fervent letter that Kenelm wrote her from college extols her mind and temperament, but says nothing about her physical beauty. Promising that he had only honourable intentions, he wrote, "you have too masculine a spirit to be frighted wth. shadowes, once you have duly surveyed and considered them". "My better Angell guided me to what I had reason to admire and love exceedingly," he averred, "And now in you, he sheweth me so many excellencies that I never was acquainted wth. before any woman, as I sweare Madame all former love wth. me hath bin but like an apprenticeshippe to teach me how to love and value you as I should." Writing after her death in 1633, his descriptions of Venetia lingered on her intellect, personality, and piety. In a letter to his brother, he wrote, "Indeed the greatnesse of her minde was beyond what ever I knew in any woman or man." "I have many times with exceeding delight heard her discourse seriously of other persons humors [sic] and actions," he recalled, "upon which she would make admirable strong observations, and hath often times made me acknowledge as undenyeable such things as, till she opened my understanding in them, I never fell to the consideration of, though I had had longer knowledge and more familiaritie with the parties than she had". In short, as he wrote defending her to a disapproving friend in Italy, "she was the best wife, the bravest woman, had the vertuousest and single soule and edified the most by her fine examples, of any woman that I have conversed with". Venetia herself does not seem to have been particularly vain, either. She was a model wife during her marriage (according to seventeenth-century expectations, of course). John Aubrey, who delighted in malicious gossip, wrote that she "carried her selfe blamelessly".

In fact, Venetia depended on her beauty for only a brief

period in her life, although admittedly it was a pretty scandalous period. One thing to be said about Kenelm and Venetia is that their relationship was always intense. They met when they were teenagers. He was three years younger than she (fourteen to her seventeen), but he was spectacularly smart and charming. It probably did not hurt that he was also tall and strong and stunningly good looking. Kenelm's mother disapproved: Venetia's family, the Stanleys, was more distinguished than the Digbys, but they did not have much money. Considering that Kenelm's father, Sir Everard Digby had been hanged, drawn, and quartered for conspiring to blow up King James and Parliament when Kenelm was three, it was pretty outrageous to say that Venetia Stanley's family was not good enough.

Like many teenagers, Kenelm found ways to get around his mother's disapproval. He wooed Venetia in person when possible and by passionate letters otherwise. She was not an easy catch by any means. As he put it later, there was no lover who "ever labored wth more passion in the gaining of [his beloved], nor mett wth greater difficulties and oppositions". Eventually, however, Venetia fell in love with him, and when he was seventeen, he persuaded her to exchange promises that they would not marry anyone else (or be unfaithful to the other, for that matter). It was a huge commitment for Venetia. She was at that time of life in which a young woman did her best to marry a titled, affluent man – preferably also young, handsome, and of good character – and there she was, promising Kenelm that she would not marry anyone else. She must have been madly in love to make such a promise, because Kenelm was about to embark on the grand tour. The grand tour was a standard event for young seventeenth-century noblemen. With some kind of adult supervision – a relative, a family friend, a tutor – a teenager like Kenelm would travel around Europe, visiting the sights, cities, and courts to gain polish, including fluency in several languages. It usually took at least two years, and a lot could happen in two years. But Venetia made the promise, and so did Kenelm, and off he went.

Things went wrong immediately. Not too long after he reached France, the young, charming, handsome Kenelm Digby was propositioned by Marie de Medici, the dowager queen of France, who was old enough to be his mother. People who said no to Marie de Medici tended to get into big and often fatal trouble, so Kenelm faked his own death and ran for it. The plan succeeded. Everyone heard that Kenelm Digby had died in France – including Venetia. Biographers including E. W. Bligh and Joe Moshenska claim that it then took her nearly two years to learn that Kenelm was alive, but that is so unlikely as to be impossible. Admittedly, seventeenth-century long-distance communication was far worse than such communication is now, but there was still plenty of interaction between England and Europe. Kenelm was not travelling incognito once he escaped from Marie de Medici, and he interacted on the road and at every stop with people who could have delivered a letter or a message. He was in Italy for more than two years, certainly enough time to write to her. After that, it was no secret that he was in Spain with his uncle, the Earl of Bristol, negotiating the marriage of the Prince of Wales. Kenelm, now in his late teens and having a glorious time in Europe, was just not making much of an effort to communicate with her.

As for Venetia, she was no idiot: she noticed. No longer feeling bound by a promise to Kenelm, Venetia became involved with at least one nobleman (although to what degree is a bit of a mystery). The story goes that when Kenelm heard, he flung himself on the floor, howling and crying at the betrayal. Oh, to be a seventeenth-century Englishman, when one could expect women to be chaste while one cheated on them all over Europe.

Kenelm and Venetia's relationship on his return was profoundly and irrevocably shaped by his silence on the grand tour. Unsurprisingly, although accounts vary of their reunion in London after three years, most agree that it was rocky. Male biographers of Kenelm have attributed both his and her standoffishness to her bad reputation, deserved or not, but that diagnosis only makes sense if Kenelm is not held responsible for his silence during those years. If he is held responsible – as he should be – her behaviour is perfectly rational and reasonable. Furthermore, when she finally agreed to renew their acquaintance, Kenelm tried to treat her like a courtesan. She required him to treat her like Venetia, the intelligent, thoughtful woman from a good family whom he had wooed devotedly and honourably, and all but promised to marry. For two years, they fought this battle. Kenelm swung between respect and contempt. Once, he snuck into Venetia's bedroom while she was sleeping, stripped, and climbed into bed with her. She managed to get him out of the bed, no small feat considering his size, strength, and undoubted resistance. One biographer maintains that Venetia berated him so soundly that he voluntarily leaped out of the bed, vowing "never again to do his love for her a disservice by being so forward and improper". Let us be clear here. The "disservice" was not to "his love for her" but to Venetia, nor is it "forward and improper" to climb naked and uninvited into a sleeping woman's bed to have sex with her. It is attempted rape.

Reader, she married him. But on her terms. In demanding that he acknowledge who she was and what she was due, Venetia reset the terms of their relationship and caused him to fall in love with her again. Once he began courting her in earnest, she fell in love with him again, as well. Admittedly, like many husbands of their time and class, Kenelm was not unswervingly sexually faithful. Nevertheless, their relationship was far more a partnership than his sexual infidelity would suggest. After their reunion in 1624, Venetia pawned, sold, or mortgaged her possessions to help Kenelm take up a prestigious post on a vital diplomatic mission. She supported his career and helped him win the approval of James I by encouraging his privateering voyage to the Mediterranean in 1626, even though she was pregnant with their second child. Kenelm sought Venetia's opinion and advice on everything. As he wrote his sons:

> I must confesse that her excellent temper in judging and great discretion in directing all affaires that was fit for me to consult with her (and I kept non from her that concerned my self) was the greatest guide and stay that I had in all my businesses [...] [S]he hath often turned my resolutions an other way, and hath mastered me with reason...

In a letter to his brother, he reflected: "That was the maine part of our happinesse, that we knew each others thoughtes as soon as we conceived them; we knew not how to reserve [any]thing from the others knowledge."

Venetia died suddenly in her sleep on the first of May 1633. Kenelm was shattered. He never remarried and wore mourning until his own death in 1665.* He had their entwined initials embroidered on the spines of his books. He wrote letters about

* Gossip later in his life accused him of having or trying to have an affair with a woman in France, and of opening marriage negotiations with a family. Even if true, the reality remains that at a time when remarriage was common, a young (for a time), handsome, wealthy, intelligent, charming man who could have remarried, did not.

her to their sons so the boys would know something of their mother as they grew up. At the time, however, his adoration manifested itself rather more idiosyncratically. After a servant discovered her dead in bed, Kenelm took plaster casts of her feet and face, and called in artist friends to paint her as she lay there and to compose a volume of poems in her honour. He also approached her death as one of science's revolutionaries and arranged a post-mortem. He was convinced that anatomy, the most developed branch of the New Science, would find the explanation for her sudden death.

What the New Science found was that her brain had turned to sludge. The cerebral cortex is not supposed to ooze out if the top of the skull is removed, but this is precisely what Venetia Digby's did. "When they came to open the head," Kenelm reported, "they found the braine much putrifyed and corrupted: all the cerebellum was rotten, and retained not the forme of braine but was mere pus and corrupted matter." Biographers have been rather casual about this bizarre finding; Moshenska does not speculate as to cause, and Michael Foster simply says that she suffered a "cerebral haemorrhage" overnight.

From a medical perspective, there is sufficient evidence to explain not only why her brain melted but also why she drank viper wine. What killed Venetia Digby? A stroke, but not in her sleep that morning. As a research team explained in 2021, "Scientists have known for years that the brain liquefies after a stroke. If cut off from blood and oxygen for a long enough period, a portion of the brain will die, slowly morphing from a hard, rubbery substance into liquid goop." The process is called "liquefactive necrosis" and that kind of necrotic (dead) tissue is toxic to adjoining tissue. In other words, liquefactive necrosis is internally contagious. Although the brain creates a barrier between damaged and healthy tissue after a stroke, toxins still slowly escape to kill off more and more healthy brain cells until, as researchers explain, "Liquefied brain tissue eventually will result in an empty cavity in which healthy brain tissue once existed." Precisely what witnesses observed at Venetia Digby's autopsy. How long does it take from the initial damage – that is, the stroke – to reach this stage? It can take months. In fact, the doctors performing Venetia's autopsy told Kenelm that the "decay they found there was not the worke of short time", and that "though the braine be the cause of sensation through the whole bodie, yet it hath none in it selfe", so she would not have felt it decaying. A stroke followed by liquefactive necrosis explains why Kenelm would have been giving her viper wine every day for a protracted period of time, although the exact period of time has not been established. As the dead brain tissue poisoned Venetia's remaining healthy neurons, she became increasingly debilitated. The viper wine was administered to revive her sinking "animal spirits". Eventually, the liquefying necrosis killed her, leaving the cerebral sludge that poured out when her skull was opened.

So that takes care of *why* Venetia drank viper wine every day, and why she drank *viper wine*. What about the fact that it was *Kenelm* who took care of the medication? The answer to this question is one of the keys to the Scientific Revolution's success at overthrowing domestic medication. Until the Scientific Revolution, food and medication were grouped under one heading because they were both made from recipes. They shared processes, equipment, materials, and the space in which they were made. Contributors to the Scientific Revolution like Sir Kenelm Digby changed the entire concept of medication, distinguishing it from food and claiming it for themselves. Where food and medication had belonged together because they shared processes, by the middle of the eighteenth century, food and medication belonged apart because they were different products. Kenelm and Venetia Digby's marriage – their division of tasks, of equipment, of knowledge, of space – shows that transformation as it took place.

When it came to domestic responsibilities, Venetia was very much in charge. As her appreciative husband put it, "She was exceeding careful and vigilant over the domestike affaires of her family, and ordered them very wisely." She gave orders the night before as to what she wanted the servants to do, and before she dressed or breakfasted in the morning, she went through the house to make sure everything was in order. Kenelm, unlike many men of his class and time, understood the difficult realities of that work. "To performe this part as it ought to be, requireth great sufficiency and stayednesse," he told his sons, "for although few of these affaires seeme to be very weighty, yet they are so many and so troublesome and of such several natures that all together they are very cumbersome and pressing."

Like other women, Venetia Digby had her own recipes. Although her recipe book has not been found, her recipes appear in books written by others. For one unidentified friend, "Ven. Stanley" provided a recipe for "A Water to Cleare Hands & Face":

Take a quart of fair water, a pint of white wine the juice of 4. Lemons, put therein Beanblossoms, white lilly Blossoms, Elder blossomes, of each one handful, put to it 4 whole Daizy roots 4 marsh mallow roots, 2 or 3 Branches of wild tansy as much ffuemitorie the weight of 2^d in Camphire, put all these in an Earthen pott, sett it in warm Ashes all night in the morning strain it through a peice [sic] of white Cotton, put it in a narrow mouthed glasse, sett it in the heat of the sea[?] 3 or 4 dayes, if there bee any pimples or rednes in the face take a quantitie of water in the glasse & steep the white dung of an hen in it one night then strame [*sic*] it through a Cloth, wash your face wth this morning & night, if your hands, add three or 4 bruised Almonds this is a most excellent water. prob.

The attribution to "Ven. Stanley" rather than "Ven. Digby" shows that she had the recipe before she was married, presumably from another woman. Recipes for cosmetics were common in recipe books. Venetia's recipe for "A Water to Cleare Hands & Face" is on the same page with "To make Breath sweet" and "A Powder for teeth", for example, and the book also contains recipes for removing smallpox scars, freckles, and sunburn. In her recipe book, Lady Sedley collected treatments for "a redd face", freckles, thinning hair, and yellowing teeth (herb and wine toothpaste).

Venetia was in charge of the household, which made them partners: Kenelm performed experiments during their marriage and she made all his work possible. For one thing, he had full access to the kitchen and anything else domestic he needed. For another, he talked with her about whatever he was thinking. Venetia's role in Kenelm's scientific endeavours and the domestic dimension of his work have been overlooked both by writers interested in Kenelm and by writers interested in the development of science and medicine. Although Sir Kenelm Digby is often recognized as an important player in the beginning of the Scientific Revolution in England, his community is described and defined as a group of men performing intellectual and experimental investigations: for example, the political theorist Thomas Hobbes, philosopher René Descartes, chemist Robert Boyle, and mathematicians Marin Mersenne and Pierre de Fermat. Kenelm was one of the earliest members of the Royal Society. Even before Charles II gave it a royal charter in 1662, when it was still a group of like-minded male investigators, Kenelm presented his own research to the other members in a paper entitled *A Discourse Concerning the Vegetation of Plants*. He had given papers all over Europe by then, starting as a young man while visiting Italy.

In reality, Kenelm's scientific circle was not all male, not all academic, and not based on the one-way transmission of knowledge from scientist to audience. After Venetia died, Kenelm became part of a group of people who shared recipes and were interested in the Scientific Revolution. Yes, that group included Robert Boyle, but it also included Boyle's sister, Katherine Jones, Lady Ranelagh, a respected chemist and political operator. The names of several other members of the group should also be familiar: Elizabeth Grey, Countess of Kent, and her sister, Alethea Howard, Countess of Arundel and Lennox. Samuel Hartlib compared the efficacy of Elizabeth Grey's "Powder" with that of Kenelm Digby's. Robert May, who cooked for Elizabeth Grey and might have had a hand in *A Choice Manual*, not only cooked for Kenelm Digby but also included him among the five dedicatees of May's own cookbook, *The Accomplisht Cook* (1660). (Kenelm was also close to John Selden, the Countess of Kent's lover.) Kenelm and the Queen Mother Henrietta Maria were good friends, sharing recipes and talking about cooking (even queens learned the domestic arts; royal they might be, women they definitely were). Apparently, he was considered an expert in making viper wine. Like a community of women, Kenelm's community shared recipes, not just for preparing food, drink, and medication but also for transmuting substances (alchemy) and understanding the components of substances and how they interacted (chemistry).

Kenelm had another connection to members of this group: like Elizabeth Grey and Alethea Howard, his papers were printed after his death without his permission. In his case, they were printed for sale at the direction of George Hartman, Kenelm's laboratory operator, who assisted him with experiments or performed them under his direction. Where the Countess of Kent's recipe book, *A Choice Manual*, was printed as a recipe book, however, the printed books with Sir Kenelm Digby's name are more amphibious. In some ways they look like recipe books, but in others they definitely do not. They have indexes or tables of contents, for example, they print recipes one after the other and only sometimes in logical order or logical grouping, and they are divided according to function – recipes for candying in one section, for medication in another.

Furthermore, like many housewives, Kenelm collected multiple recipes for the same substance. Ninety-eight of the first 103 recipes in *The Closet of the Eminently Learned Sir Kenelme Digbie Kt. Opened* (1669) are for meath and metheglin (variations of mead), fermented medicinal drinks used when someone was "feeling poorly", so to speak. Many of the culinary recipes in his books also use conventional recipe language. For instance, in *Two Treatises* (1669), three pudding recipes in a row – "To make excellent Black-puddings", "To make White Puddings", and "To make an excellent Pudding" – begin with "Take", and the recipe for white pudding warns that the "small-guts [...] are to be cleansed in the Ordinary manner; and filled very lankley; for they will swell much in the boiling, and break if they be too full".

Also like an ordinary recipe book, many of the recipes came from people Kenelm knew, such as "Sweet-meats of my Lady Windebanks" and "Mr Webb's Ale and Bragot". Several of those people served in the court of Henrietta Maria, the queen mother, where Kenelm was an officer starting in the 1640s. Her Majesty shared her own recipes – "White Marmulate, the Queen's Way", for instance – and gave advice on roasting meat: "The Queen useth to baste such meat with yolks of fresh Eggs beaten thin, which continue to do all the while it is rosting." Kenelm also recorded recipes for dishes she was served, such as "Portugal Broth, as it was made for the Queen". The recipe for "Portuguez Eggs" begins: "The way that the Countess de Penalva makes the Prtuguez [sic] Eggs for the Queen, is this."

Language like "the way that" and "was made thus" is not like the usual instructions beginning with "Take" or "Put". Instead, such language elevates the process to be equally important or more important than the product. "Pressis Nourissant" begins "The Queen Mothers Pressis was thus made," for example, and "The Queens ordinary Bouillon de santé in a morning, was thus." This rhetorical difference is not minor. Once recognized, Kenelm's interest in process appears everywhere, so much so that the books seem less like recipe books and far more like laboratory notebooks. Women's recipes begin with directions, almost always the word "Take". Kenelm's are not so regular. Many of his recipes combine instructions with source when they begin: "My Lady of Monmouth boileth a Capon with white broth thus," for example, or "My Lady Homeby makes her quick fine Mustard thus..."

When Kenelm persuaded "Master Webbe, who maketh the Kings Meathe" to show him how the King's Meathe was made, he recorded not just the recipe but also the entire event as a narrative. "The first of Septemb. 1663. Mr Webb came to my House to make some for Me," he began, adding commentary along the way: "I am not satisfied, whether he did not put a spoonful of fine white good Mustard into his Barm,* before he brought it hither, (for he took a pretext to look out some pure clean white barm) but he protested, there was nothing mingled with the barm, yet I am in doubt." Kenelm must have suspected that Mr Webb was not going to give him the actual recipe, hence the invitation to his house so he could watch the procedure (and even then Mr Webb appears to have been less than straightforward). One of the clearest statements of Kenelm's interest in process is the recipe for "Slipp-coat Cheese", which begins "Master Philipps his *Method* and *proportions* in making slippe-coat Cheese, are these" (emphasis added).

Kenelm's own process is often part of the recipes. There is his eagle-eyed account of watching Mr Webb make meath, of course. The rather smug recipe in *Two Treatises* for "Hydromel as I made it weak for the Queen Mother" concludes, "Thus was the Hydromel made that I gave the Queen, which was exceedingly liked by every body." (Hydromel was a kind of laxative that used water and honey. It suggests a considerable level of intimacy, not to mention gastric distress, that "every body" used Sir Kenelm Digby's hydromel at some point, although Kenelm was also prone to bragging.) When he records "My Lady Bellassises Meath", the recipe begins: "The way of making is thus. She boileth the honey with Spring-water, as I do, till it be cleer..." Kenelm himself shows up quite a lot in *Two Treatises*. Explaining one of his recipes, he announces:

> I have found it admirable for the Brain, the Eye-sight, the Heart, the Stomach, and all languishes Diseases and decays

* Barm is the layer of yeast forming at the top of malt drinks while they ferment.

of Nature, and causeth a little gentle breathing, scarce amounting to swear: When you take it in a morning, it gives you a wonderful severity of brain and cheerfulness of humour in languishing Diseases.

Every now and then, his humour appears: "Into this liquor put two ounces of good old Venice Treacle." Good old Venice Treacle indeed.

While the contents of the books printed and sold from Kenelm Digby's papers make classification challenging, the titles reveal how experiments and recipes were being used by science's revolutionaries. Hartman printed the first book from Kenelm's papers in 1668, with the permission of Kenelm and Venetia's sole surviving child, John Digby, and crediting himself ("GH") as translator. The book was entitled *Choice and Experimented Receipts in Physic and Chirurgery: as also Cordial and Distilled Waters and Spirits, Perfumes, and other Curiosities, Collected by the Honourable and truly Learned Sir Kenelm Digby Kt.* Since the title announces that it contains recipes, it would seem to be a recipe book. On the other hand, the recipes are "experimented", which suggests that there is something scientific going on. It was usual for recipes in "Physick", that is, recipes for medicinal treatments, to appear in recipe books. However, recipes for "Chirurgery", the physical aspect of medical care like amputation and excising growth, were not usual to recipe books. Furthermore, there were plenty of printed and private books of recipes dedicated solely to medicinal treatments, and "cordials" and "waters" could have medicinal value. On the other hand, *Choice and Experimented Receipts* also has instructions for making perfumes and "Distilled Spirits" (not to mention that intriguing category, "other Curiosities"), which are for domestic use. The title does not definitively classify the book, nor do the contents. It is a platypus: definitely something, but not like anything.

The next year, George Hartman had another volume of Kenelm's records printed for sale: *The Closet of the Eminently Learned Sir Kenelme Digbie Kt Opened: Whereby is Discovered Several ways for making of Metheglin, Sider, Cherry-Wine, etc. Together with Excellent Directions for Cookery, as also for Preserving, Conserving, Candying, etc.* Readers of the previous chapter may notice the similarities of this title with the titles of other recipe books, although Hartman eschewed the titillating innuendo. *The Closet of the Eminently Learned Sir Kenelme Digbie* does not use the word "recipe" (or "receipt") anywhere on its title page. Although the book is full of recipes, it avoids that term. Instead, Hartman replaced the word "receipts" with the phrases "Several ways for making" and "Excellent Directions for Cookery". Here is a recipe book that is not a recipe book, even though it contains only recipes for food and drink.

Hartman saw a second book into print in 1669: *Two Treatises, By the Honourable and truly Learned Sir Kenelm Digby Knight. The one, Of Choice and Experimented Receipts in Physic and Chirurgery; as also Cordial and Distilled Waters and Spirits, Perfumes, and other Curiosities. The other, Of Cookery, With several ways for Making of Metheglin, Sider, Cherry-Wine, etc. Together with Excellent Directions for Preserving, Conserving, Candying, etc.* Hartman combined both books into a single one, creating something that looks like a standard recipe book with medicines in one part and foodstuffs in another. The trick is in the eye-glazing subtitles. They underscore the distinction between the recipes for "Physic and Chirurgery" on one hand and those for "Cookery" on the other. The title for the combined book, *Two Treatises*, emphasizes that two complete works are bound together; it is not one whole work with two sections. *Two Treatises* is itself an odd title. Calling a collection of recipes a "treatise" makes a claim for coherence and unity, not to mention a purpose and a thesis, that a recipe book does not have. The title transforms the collection of recipes into something from the intellectual and traditionally male aspect of culture. Using the title *Two Treatises* is also a clever marketing strategy, as it was the title of a well-known publication of Kenelm's from the 1640s that addressed theological issues. Hartman's *Two Treatises* from 1669 is another amphibian, neither one genre nor another but something in the middle of its development.

The titles to Kenelm's posthumous publications reveal that there was a great self-consciousness about how to classify his work. Kenelm – a man – could have recipes for medicinal substances for use in "Physick and Chirurgery", but when it came to edibles and drinkables, he had "ways" and "directions". It was women who had recipes for cider, cherry wine, and roasted venison. The publication of Kenelm's recipes reflects a growing distinction between medication and comestible. The recipes are in separate books, and the language for titling and describing the contents of each book is different.

Two Treatises was never reprinted. After 1669, the two parts were printed and sold separately. *Choice and Experimented Receipts* (primarily medicinal treatments) was reissued in 1675, and *The Closet of the Eminently Learned Sir Kenelme Digbie* (primarily cookery) came out in 1671 and 1677. In 1682, Hartman again made use of his late employer's papers to produce *A Choice Collection of Rare Secrets and Experiments in Philosophy, as also Rare and unheard-of Medicines, Menstruums, and Alkahests; with the True Secret of Volatilizing the fixt Salt of Tartar. Collected And Experimented by the*

Honourable and truly Learned Sir Kenelm Digby, Kt. The title page also announced that this collection was "Hitherto kept Secret since his Decease, but now Published for the good and benefit of the Publick". It was printed again that year as *A Choice Collection of Rare Chymical Secrets and Experiments*, and in 1683 as *Chymical Secrets and Rare Experiments in Philosophy*. The evolution of the titles for Kenelm's books from 1669 to 1683 reveal that big changes were taking place in the composition of what was considered the best medication – chemical medication was growing in prestige and replacing organic medication. That is a subject for Chapter 6, however; I am pointing it out here only in passing. For now what is of interest is the change from "Choice and Experimented Receipts" in 1669 to "A Choice Collection of Rare Chymical Secrets and Experiments" in 1682. "Choice" has to do with the collection, not the recipes; the adjective "Experimented" has become the noun "Experiments".

Kenelm's posthumous works offer a timeline showing how the proponents of the Scientific Revolution drove a wedge between food and medication, gendering each side of the divide. This separation contributed to reconceptualizing the body and health. Until the middle of the seventeenth century, men like Kenelm and Francis Bacon conducted experiments with food as food, but as the Scientific Revolution progressed food was increasingly excised from scientific investigation. Before about 1650, books about maintaining health often discussed diet and sometimes referred to it explicitly in their titles. Tobias Whitaker may have been prompted by wishful thinking to pen *The Tree of Humane Life, or, The Bloud of the Grape. Proving the Possibilitie of Maintaining Humane Life by the Use of Wine* in 1638, but others were more practical. Sixteenth-century titles, for example, included *An Introduction into Phisycke, with an Universal Dyet* (1545) by Christopher Langton and *The Olde Mans Dietarie* (1586) by Thomas Newton.

At the same time that "physic" was evolving into a category that did not include diet, domestic medicine was evolving into a category that was primarily about food. The phrase "kitchen physic" changed meanings between 1650 and 1740. Initially, the term referred to workaday, reliable medication. In *A Pisgah-sight of Palestine* from 1650, Thomas Fuller wished upon a frail, older man "good kitchin-Physick, carefull attendance, and serious meditation on his latter end". In 1677, however, the distinguished Scottish physician Matthew Mackaile explained in *A Short Treatise, Concerning the Use of Mace, in Meat, or Drink, and Medicine* that his work was "particularly recommended unto such of the Female Sex, as are most studious, only of the Diecteticall part of Medicin (commonly called Kitchin Physick) it being chiefly of that nature, and most properly belonging unto them". Thomas Tryon capitalized on the success of his first book, *The Good Housewife made a Doctor* (1692) with *A Pocket-Companion* (1693), which contained extracts from *The Good Housewife* and claimed to provide "A plain Way of Nature's own Prescribing, to Cure many Diseases in Men, Women and Children, by Kitchen-Physick only". By the end of the 1730s, "kitchen physic" meant a healthy diet, either for preserving health or helping a patient regain strength after illness. In his manual for midwives, *The Midwife's Companion* (1737), Henry Bracken prescribed his own medication to stop heavy bleeding after childbirth and recommended "Kitchen Physick" once the bleeding stopped, to help a woman get back to strength. Two years later, William Dover asserted that "Kitchin Physick is the best medicine" for staying healthy. Viper wine when you are ill, chicken soup to keep you well.

At the same time that these changes were taking place,

commercial household manuals were shrinking women's responsibility for medication. Early books aimed at women mentioned both comestibles and medication in the main title. There was *The Good Hous-Wives Treasurie. Beeing a Verye Neccessarie Booke [of] the Dressing of Meates. Also Sundrie Medicines* in 1588 and *A Closet for Ladies and Gentlewomen, or, The Art of Preserving, Conserving, and Candying. Also Divers Soveraigne Medicines and Salves*, printed nine times between 1608 and 1636, an impressive sales record. John Partridge, whose books ran to many editions apiece, sold *The Treasurie of Commodious Conceits, & Hidden Secrets. And May be Called, the Huswives Closet, of Healthfull Provusion* in 1573 and *The Widowes Treasure, Plentifully Furnished with Secretes in Phisicke and Chirurgery for the Health and Pleasure of mankinde. Hereunto are Adioyned, Sundry Pretie Practises and Conclusions of Cookerie* in 1586. *The Treasurie of Commodious Secrets* was reprinted until 1637 and the more titillatingly entitled *Widowes Treasure* was reprinted until 1639, when it competed with a hot new title, *The Ladies Cabinet Opened, Wherein is Found Hidden Several Experiments in Preserving and Conserving, Physicke, and Surgery, Cookery, and Huswifery.*

Compare these manuals for women to those at the end of the period. In 1723, John Nott, a celebrity chef of his day, put out *The Cook's and Confectioner's Dictionary: Or, the Accomplish'd Housewife's Companion* for the "Use of you *British Housewives*, who would distinguish yourselves by your well ordering the Provisions of your own Families". Nott also considered it a "necessary Companion also for Cooks, &c, in Taverns, Eating-Houses, and publick Inns, and not an unnecessary one, for those who have the ordering of noble tables". About a decade later, another celebrity chef, John Middleton, produced *Five hundred new receipts in cookery, confectionary, pastry, preserving, conserving, pickling; and the several branches of these arts necessary to be known by all good housewives*. Neither book has any recipes for "physick". Charles Carter's *The Compleat City and Country Cook: or, Accomplish'd Housewife*, a sumptuous collection of "Several hundred of the most approv'd receipts in Cookery" and furnished with forty-nine illustrations, was in its second edition in 1734, and mentioned almost as an afterthought that it also had "two hundred" recipes for standard family injuries and ailments, including several for treating "the Bite of a Mad Dog". Just a few years later, "Mrs Sarah Harrison, of Devonshire" issued an expanded edition of *The House-keeper's Pocket-Book And Compleat Family Cook* (1739). Although it claimed 700 culinary recipes, *The House-keeper's Pocket-Book* only contained "many" aids for treating family health problems. Furthermore, these aids were "excellent Prescriptions" that had been taken from other physicians' books rather than "receipts" or "recipes". Where later printed housekeeping books mentioned recipes for physic, those recipes were almost always in the back, in small print. Some recipe books divided food recipes from medication recipes, and some recorded the latter in the second part of the book. The consistency of this placement and the very literal shrinking of the space in the books given to those recipes attests to the removal of medication from women's responsibilities and the emphasis on edibles and drinkables instead.

The change of space in housekeeping manuals reflected a change in domestic space. As women's responsibilities changed, so did the places that belonged to them. The kitchen and the garden were the most obvious domestic spaces affected, but the whole notion of "domestic" was impacted: after all, if "making and administering medication" belong to others, they no longer belong to "domestic". As medication and food went their separate ways – or were perceived to be separate entities – the kitchen lost its function as a place for making medicinal goods.

The changing physical architecture of Digby households epitomizes this evolution. Initially, Kenelm began experimenting on food in the kitchen of the house in Charterhouse Square in Clerkenwell, where he and Venetia lived with their children. As a widower, Kenelm was more peripatetic, and his children usually stayed with his mother or at school. Wherever he lived, however, he created experimental space. Immediately after Venetia's death, Kenelm moved to Oxford and maintained a laboratory in his lodgings. The inventory shows that it was stocked with, among other things, several furnaces, a "Reverberating Calcining Oven", numerous glass bottles, tongs, and grinding implements. It also had "His New Oven to bake pies in". When Kenelm lived in Paris, he also maintained a laboratory, and when he lived in Covent Garden in London at the end of his life, his house had two kitchens, one for experiments and one for preparing food. (His friend Alethea Howard, who was exponentially richer, built herself a whole separate house in London for conducting experiments and making medications from recipes.) Kenelm even established space for performing experiments when he was a prisoner in the 1640s.

Over his lifetime, his household staff reflected the redistribution of space that came from the growing division between medication and experiment on one side and food on the other. Initially, he and Venetia employed a cook, who occasionally assisted him or got out of the way. In Oxford, he had laboratory operators to help him conduct experiments who were separate from his household staff, including Hans Hunneades.

When Kenelm returned to London, he employed a cook for the cooking and a laboratory operator to assist with the experiments, Anne and George Hangmaster, respectively. George Hartman also worked with Kenelm in this house. The point is, at the start of the seventeenth century, Kenelm like so many other experimenters was highly dependent on domestic space, specifically female domestic space, for room and equipment to undertake his experiments. By the end of the seventeenth century, kitchens and laboratories were beginning to be distinct spaces, the one belonging to women like Venetia Digby (had she survived) and Anne Hangmaster, the other belonging to men like Kenelm Digby, George Hangmaster, and George Hartman.

For men interested in experimental science, this was a gain: the laboratory was a whole new space belonging to them. For women, whether they were interested in the New Science or in keeping intact the lives and limbs of their families, it was not. As women lost their responsibility for medication, their role in the household – and therefore their authority – shrank. And as the kitchen took on a more limited function, its role in the household also shrank. The concept of "domestic" shrank accordingly. Without medication, "domestic" lost a whole realm of responsibility and its outward, community orientation. As the idea of "domestic" contracted, family life also contracted and communities increasingly consisted of discrete units banded together and less and less an organic, syncretic whole.

If the kitchen was one front in the battle to separate food and medication, then the garden was another. Medication was almost entirely organic at the start of the Scientific Revolution, and many if not most of its ingredients were grown at home, just like most and often all ingredients for meals. The garden was vital to the survival of the household, which made it domestic space and women's responsibility. As "FB" explained in *The Office of the Good House-wife* in 1672, the housewife is responsible for the "Cure & Charge of the Families health"; therefore the "good House-wife must have a good share in the oversight" of the garden – by which "FB" actually meant all of the oversight. Even queens were expected to manifest responsibility for gardens. In the middle of the seventeenth century, the Queen Mother Henrietta Maria had a garden planted for her that included artichokes, a fashionable new plant.

For centuries, every household that had the space for it, had a garden. Maps of grand estates show that they usually kept their kitchen gardens out of sight, as the great botanist John Parkinson advised. In urban neighbourhoods, people grew household necessities in the areas behind and between houses. Seventeenth-century maps of cities such as London and Bristol show that in many neighbourhoods, gardens were standard; guilds maintained gardens beside their meeting halls and, in London, garden sheds were so common that they turn up frequently in the records for crimes such as burglary or selling illegal substances.

Most kitchen gardens were outside of cities, however. Every inch of a kitchen garden's space was used, including the walls, and every plant had a purpose: to attract and feed bees, to feed people, to supply fragrant floor coverings, to make soap or other household goods, to treat ailments, illness, and injury in humans and animals, and so on. John Smith, an early agricultural scientist, calculated that the ideal small estate should include a three-acre "garden of herbs and roots" divided equally between a "Kitchen Garden" and a "physick-garden". Caring for the home garden demanded from the woman of the house a tremendous amount of knowledge, skill, labour, and time. Unsurprisingly, housewives maintained their own storehouse of knowledge, seeds, and tools. As they did with recipes, they shared cuttings, seeds, gardening tips, and so on.

Like cookery books and housekeeping manuals, printed gardening books written specifically for women began to appear in the seventeenth century. The very first gardening book for women, *The Country Housewife's Garden* by William Lawson, first went on sale in 1618 with a companion volume for men, *A New Orchard and Garden*. It sold outstandingly well, particularly after Gervase Markham acquired it and started churning out editions. Other gardening books followed. "FB" devoted fourteen pages of *The Office of the Good House-wife* to caring for the kitchen garden, and another seven to "the Garden of Pleasure" (plants notable for colour and scent). *The Accomplisht Lady's Delight in Preserving, Physick, Beautifying, and Cookery* (1675), sometimes mistakenly attributed to the cookbook author Hannah Woolley, included "The Lady's Diversion in Her Garden", which provided instructions for maintaining the family gardens and ornamental greenery around the house and concluded with a month-by-month planting and garden maintenance schedule. *The Office of the Good House-wife* also provided a month-by-month planting schedule and detailed instructions for beekeeping because, "If the greatest part of the profit of a Farm depend upon the keeping of Cattle: I dare be bold to affirm, that the fruitfulest thing that can be kept about a Countrey-house, is Bees." The *Woman's Almanack, for the Year 1694* begins with a monthly calendar for maintaining the orchard and gardens, and caring for bees.

Lawson's vocabulary in *The Country Housewife's Garden* reflects organizational thinking before the Scientific Revolution. He called everything a herb: artichokes, chamomile, coriander, featherfew (or feverfew), fennel, lilies, lovage, parsnips, poppies, and so on. This general definition makes sense when you think of the root (association intended) of the word "herbaceous". Lawson's instructions organized the housewife's garden based on the frequency of harvest and replanting: the "durable" garden bloomed every year, while "that which is [for] your Kitchens use, must yield daily roots". The ingredients for food and medication were grown together because the two products belonged together, but the truth was that distinguishing between an edible and a medicinal plant was often impossible. Some plants promoted health, others had a medicinal effect but were eaten at meals – were they medicinal or culinary? Both.

By the middle of the eighteenth century, however, gardening instruction for women had little to say about maintaining medicinal plants. Small wonder. In 1651, Nicolas de Bonnefons wrote *Le Jardinier François* (*The French Gardener*) for a female audience, and his conception of the garden was female. When John Evelyn, a member of Samuel Hartlib's circle and a founding member of the Royal Society, translated *Le Jardinier François* into English in 1658, however, he re-gendered it male. Over the next few decades, gardening books increasingly addressed male readers and increasingly dismissed women's contributions. In *The Art of Gardening, Improv'd* (1717), John Evelyn's son Charles stated that the "curious Part of Gardening in general, has always been an Amusement chosen by the greatest of Men". The "Management of the Flower-Garden in particular, is oftentimes the Diversion of the Ladies", Charles explained, but only "where the Gardens are not very extensive, and the Inspection thereof doth not take up too much of their Time". In his view, "the fair Sex" needed "Encouragement" to take up gardening. Although Henry Stevenson addresses "Gentlemen and Ladies" at the very start of *The Young Gardener's Director* (1716), every other page, including the title page, uses male pronouns; he states that he designed the book for "young men" to carry around in a pocket.

By the late 1720s, gardening books were directed at men. Stephen Switzer's *The Practical Kitchen Gardener* (1727) never considered that a woman might be involved in maintaining the kitchen garden. His and other authors' texts reinforced women's exclusion from gardening by requiring the kind of erudition only available to men, such as knowledge of botanical experts, ancient writers, and Latin. "I can't help considering a good Gardener both as a philosopher and a politician," Switzer wrote. Richard Bradley's *New Improvements of Planting and Gardening* (1717) was *Both Philosophical and Practical* and therefore only comprehensible to men – women, of course, did not have the education (or the brains) to understand it. By 1739, the incomparably named Batty Langley was arguing that the "Pleasure of a Garden depends on the variety of its Parts" and asserting that "all the most useful elements of Geometry" were "necessary to be understood by every good Gardener", regardless of whether the garden was for aesthetic enjoyment, growing fruit, or supplying the kitchen.

European imperialism was vital to efforts to assign medication and physic gardens to men – another way in which imperialism and the Scientific Revolution conspired together. As Europeans arrived and strengthened their presence in Asia, sub-Saharan Africa, the Americas, and the Caribbean, they encountered flora that they had never seen before. These plants became an object of fascination and a commodity. The potential for new medicinal treatments was exciting and the desire to have the latest exotic item from distant locales is only human. Boasting the most extensive collection of exotic plants became fashionable among those with the money and space. Plants became commodities; collecting became a serious international business. The Dutch established nurseries in South Africa to feed the collecting craze for plants gathered from around the Indian Ocean. (In 1668, the Dutch cemented their hold on that region by obtaining the last of the Spice Islands in exchange for an island at the mouth of the Hudson River. History had the last laugh.) The English finally caught up with the Dutch in terms of horticulture in 1681, when George London and Henry Wise founded London and Wise's, the first plant nursery in England. As Richard Pulteney wrote in *Historical and Biographical Sketches of the Progress of Botany in England: from its origin to the introduction of the Linnæan system* (1790), "The growing commerce of the nation, the more frequent intercourse with Holland, where immense collections from the Dutch colonies had been made, rendered these gratifications more easily attainable than before." The popular frenzy for collectable plants was also partly inspired by Mary II. Before returning to the British Isles as queen in 1688, she and her husband, William of Orange, had collected rare specimens from around the world. When she moved back to the land of her birth, in addition to a Dutch husband, she brought the latest horticultural technology. Thanks to the Scientific Revolution, starting around the middle of the seventeenth century, people also started collecting plants for intellectual or research purposes. Others, like John Goodyer, William Coys of Stubbers, Edward Morgan, and William Sherard, a diplomat, had the time, money, and interest to establish impressive gardens and often amass equally impressive libraries about horticulture. By the end of his life, Goodyer had learned the medicinal properties of many plants and was providing "simples" for people in his neighbourhood. In other words, he had turned into a housewife.

Men like Goodyer and Morgan and women like Mary II were not scientists. The Scientific Revolution not only encouraged the deliberate collection of plants and knowledge but also developed a methodical approach to the botanical world. Science's revolutionaries wanted to organize and make use of these new plants, this new knowledge, and this expanded view of nature. Pulteney approvingly noted that domestic peace and international trade fostered the interest in collecting rare plants, and "from all these happy coincidences, science in general reaped great benefit". Botanists were part of the New Science's active, international network through which knowledge, ideas, equipment, results, and even seeds and plants were exchanged. They endlessly catalogued and recatalogued as new herbaceous life became known to Europeans. John Ray developed one way of classifying plants in the late seventeenth century; Carl Linnaeus developed his Linnaean taxonomic system in the early 1730s.

Collectors and botanists formed clubs that met regularly at coffee houses such as the Temple Coffee House and the Rainbow Coffee House in London. Members corresponded, exchanged seeds and samples, commissioned each other to find rarities, and wrote papers for scientific journals like the Royal Society's *Philosophical Transactions*. Even Sir Kenelm Digby, who was not a plant collector, gave a paper on *The Vegetation of Plants*. Unsurprisingly, with the very rare exception, these people were men. Furthermore, many of the most dedicated investigators and collectors of new flora were Fellows of the Royal Society, the organization founded in 1662 to promote the New Science.

Many of them were also medical men. Physicians Sir Hans Sloane, Alexander Stuart, and Richard Mead, for example, and well-known apothecaries like Isaac Rand, Samuel Doody, and James Petiver all were Fellows. Petiver also made a fine living selling seeds and samples from his collection. Another Fellow of the Royal Society, James Sherard, trained as an apothecary and later also received a medical degree; his garden of exotic and medicinal plants at his estate in Eltham was famous. The private plant collector William Sherard was his brother. When William died, he endowed a professorship in botany at the University of Oxford, with the condition that the professor should be chosen by the RCP. James Douglas, the most prominent obstetrician of his day and a central figure in Chapter 6, published papers on plants in *Philosophical Transactions*. He also read papers to the Royal Society on the curative properties of different plants and trees, the discovery of a new kind of narcissus, and the cultivation of "English saffron", an ingredient in organic medications.

A notable exception to this boys' club was Mary Somerset, Duchess of Beaufort, whose spectacular wealth, intelligence, and interest made her a vital contributor to the development of botany in the late seventeenth century. She was so important that Richard Pulteney, who neglected to mention Queen Mary's work at Hampton Court, named Somerset among the most significant early botanists. She obtained thousands of cuttings, plants, and seeds from across the globe; what she could not get through personal connections, she commissioned from agents. One shipment from Barbados had to be divided and transported on five different ships because the specimen containers were so large and so numerous. The Duchess of Beaufort was committed to the Scientific Revolution in botany. She received *Philosophical Transactions* to keep up with the latest discoveries. She identified, classified, and catalogued everything she collected; kept illustrations of her samples; and hired her own assistants to help with her work. She also exchanged ideas, materials, and the results of experiments – including her own – with men including Sloane, Petiver, Ray, and William Sherard.

The duchess became interested in botany after learning of

Joseph Glanvill's recommendation to study nature through the scientific method as a way of treating depression. She exemplifies the convergence of the woman's domestic responsibility to create treatments for family ailments, the idea that medication is organic, and the New Science. Unfortunately, she is an exception who proves the rule: the New Science's male revolutionaries determinedly established that studying plants, medication, and medicinal plants was their business. (After her death, Charles Evelyn generously conceded that the Duchess of Beaufort "thought it no Diminution to concern herself in the directing Part of her Gardens" and called her the "greatest Example of Female Horticulture", which was a "most pleasant and agreeable Employment".) Gardens created by proponents of the Scientific Revolution were scientific gardens. These were spaces that belonged to men, were maintained by men, and served male interest in knowledge. They were not spaces whose primary function was to sustain a household through the care and feeding of its members.

Unsurprisingly, there was a simultaneous surge of academic interest in medicinal gardens. Almost the moment that Andreas Vesalius published *De Humani Corporis Fabrica* in 1543 and kicked off the Scientific Revolution, medical schools across Europe began establishing gardens of medicinal plants. The University of Pisa established Europe's first physic garden in 1544; the University of Padua followed in 1545. From there, academic medicinal or physic gardens for training physicians spread rapidly. The University of Valencia established a position for teaching medicinal compounds and a teaching garden to support the curriculum in 1567. The University of Leiden began its famous physic garden, the Hortus Botanicus (better known as the Hortus) in 1577; in France, Montpellier broke ground in 1595 and Paris in 1596.

Across the English Channel things got going a little later. John Gerrard's efforts at the end of the sixteenth century to establish physic gardens for the barber-surgeons' guild and for the University of Cambridge came to nothing. In London, the RCP aspired as early as 1587 to start one, but there are no documents proving that it existed until 1651. Edinburgh's apothecaries and College of Surgeons together founded a physic garden in 1656. As for educational institutions, the University of Oxford began its physic garden in 1621, but the other great educational institutions of the British Isles did not follow suit until after 1650. The University of Glasgow established a physic garden in 1704 and the University of Cambridge, clearly not in the vanguard, finally broke ground in 1762.

From at least as early as the late 1500s, during Elizabeth I's reign, some physicians and apothecaries kept personal gardens to supply materia medica, but it was hardly a system for training apprentices. The apothecary's apprentice had a lot in common with the daughter in a household. He had to learn what parts of what plants processed in which ways produced which medications, and he had to learn to recognize medicinal plants in a garden and in the wild. To provide a thorough, standard, reliable education in materia medica to their apprentices, in 1673 the Worshipful Society of Apothecaries leased a small plot on the Thames to use as a teaching garden. The Chelsea Physic Garden is still there, a bright, beautiful island in the middle of Chelsea.

From its beginning, the physic garden was a product and a tool of the Scientific Revolution. The apothecaries created administrative and teaching positions: the *Horti Praefectus*, who oversaw everyone and everything at the physic garden; the head gardener; and the demonstrator, who taught the apprentices. Its earliest officers were active in the New Science. James Petiver and Isaac Rand were the first two demonstrators. Rand also served as *Horti Praefectus* for nineteen years. Samuel Doody was the head gardener from 1692 to 1706. His immediate predecessor, James Watts, established a partnership between the physic garden and the Hortus at the University of Leiden that outlasted individuals and provided both institutions with invaluable seeds, samples, and information. In the scientific spirit of technological innovation, in 1680 the apothecaries built England's first greenhouse to keep alive delicate plants from foreign climes. Mary II was so impressed that she added a greenhouse to the gardens at Hampton Court to protect her exotic plants. To foster research and learning, the apothecaries also established a library for books about materia medica, botany, and gardening. Any member could use them after getting permission from the *Horti Praefectus*. And, of course, there was the collecting. Head gardeners collected almost fanatically.

Another aspect of the apothecaries' physic garden that was tied to the Scientific Revolution was the collaboration of the apothecaries and the physicians to establish and maintain it. Before 1650, the two professions competed for patients, money, and control of medication, and in truth this competition did not vanish with the end of the Civil War. Far from it. Nevertheless, significant members on both sides shared two vital interests: the Scientific Revolution and taking over medication from women. The physic garden was a powerful tool in promoting both in one go. When the Worshipful Society of Apothecaries was in danger of losing the land on which they planted their garden, the organization turned to Sir Hans Sloane for help. While buying his estate in Chelsea, Sloane purchased the physic garden's land and leased it to the apothecaries for five pounds and delivery of fifty samples to the RCP every year. The lease still runs, although the terms have changed slightly.

Interest in medicinal plants united the apothecaries and the physicians, and Sloane's arrangement eliminated competition for knowledge and materials. The relationship was reinforced by the close ties with the University of Leiden. Many of the Fellows of the Royal College had been trained at Leiden's prestigious medical school and in its spectacular Hortus, while the apothecaries and the Hortus exchanged materials and information. Long after Sloane and his colleagues – including Mead, Rand, Petiver, Sherard, Douglas, and Doody – were gone, the Chelsea Physic Garden brought together physicians and apothecaries in the scientific pursuit of botanical medicine. The collaboration and like-mindedness regarding the physic garden intensified over the late seventeenth and early eighteenth centuries. By the 1730s, physicians often attended dinners at Apothecaries' Hall and regularly were guests at the "public" herbarizing for apprentices in July. Herbarizings were monthly summer expeditions into the countryside led by master apothecaries. One important use for the guild's official barge was to transport apothecaries and their apprentices down the Thames on these occasions.

It was at the dinner following the public herbarizing in July 1734 that the RCP and the Worshipful Society of Apothecaries agreed to support another scheme promoting their superiority: Elizabeth Blackwell's *A Curious Herbal*. Blackwell (1707–58) proposed to make a beautiful, two-volume herbal for commercial sale – a luxury item, a boutique publication – that showed off the spectacular extent of apothecarial knowledge while linking it with the New Science. Her plan was brilliant and innovative: she organized the images and the information so each one and the book as a whole appeared methodical, but she also rendered the images and information visually compelling. Her proposed target was not women as a whole, but "herb women" – that is, women who competed with apothecaries by selling medicinal herbs. All those women who could afford to buy the print sets or complete volumes remained unoffended, but they were also directed towards male authority.

The apothecaries and physicians embraced Blackwell's proposal, publicly announced their support, and gave her unlimited access to the physic garden. Over the next four years, despite extreme weather, two pregnancies, the deaths of both infants, the hostility of the head gardener, the theft of her work by rival printers, and a deadbeat husband who sold all of her etched copper plates (even the unfinished ones) to settle his debts, she produced 500 etchings of plants and another 125 etched information pages. All in the service of the apothecaries and physicians' campaign against domestic medicine and the women who practised it. When she died in 1758, a group of physicians and apothecaries arranged for her burial next to Sir Hans Sloane in Old Chelsea Churchyard as a mark of respect.

Elizabeth Blackwell was also the only woman permitted in the physic garden. A woman could be brought to the garden as a guest, but they were not allowed into the garden on their own. The *ne plus ultra* of professional physic gardens, the apothecaries' garden in London was thus also the antithesis of domestic medicine's medicinal garden, which was female space. Domestic medicine's garden belonged to the home and was literally next to it; commercial medicine's garden was educational and professional space, physically separated from the home. (Elizabeth Blackwell and the head gardener lived across the street from the physic garden; they both commuted to it.) Domestic medicine's garden was female space; the professional physic garden was male space. Domestic medicine's garden was open to the household; the professional physic garden was off limits except to authorized professionals. And of course, women were allowed in domestic medicine's garden but banned from the professionals' garden.

Kenelm and Venetia Digby's married life together was short and sometimes tempestuous, but it also encapsulated the transformation that the Scientific Revolution wrought on food and medication. Initially, Kenelm's experiments and Venetia's space, the kitchen, literally and metaphorically overlapped. As Kenelm continued his experiments after Venetia's death, he increasingly separated one space – the laboratory where experiments were conducted – from the other – the kitchen where recipes were followed. And while Kenelm himself did not necessarily distinguish medication from food as cleanly in life as he appeared to in death, the posthumous publication of his work promoted and attested to a cultural distinction between the two and a re-gendering of medication from female to male.

The whole process, of course, took longer than the span of Kenelm's life and the life of editions of his notebooks, but it had succeeded by 1740. Printed housekeeping, gardening, and cookery books for women had little or nothing to say about women's role as medicine makers and everything to say about women's role as food providers. Medicine had been disentangled from food, or maybe reconceptualized, and associated with men. Although in practice plenty of women continued to grow physic gardens and make medicine to treat their families, friends, and neighbours, the cultural associations had been rearranged. Elizabeth Blackwell's *A Curious Herbal* (1737/39) glorified the scientific approach to medication, promoted its connection with physicians and apothecaries, and denigrated women as medicators. The work itself, its purpose and supporters, and its success all testify to the reconfiguration of food and medication between 1650 and 1740. Had the Digbys lived a century later, Kenelm would have had to become an expert in chicken soup, not viper wine, if he wanted to take care of his beloved Venetia.

4

Proscriptions, Prescriptions, and Poetry

What do you call a woman physician? That depends on when and where you are asking. If the year is 1674 and the place is England, the answer is "Mrs". If the year is 1675 and the place is England, the answer is "Medicatrix". At least, that is what Mary Trye would tell you. If the year is 1685 and the place is England, the answer is "Doctor" – according to Jane Barker. By the middle of the eighteenth century, the answer is "an impossibility" or "a freak of nature".

A rose by any other name might smell as sweet, but titles matter, and starting around 1650, the meaning of certain medical titles changed significantly. What did it mean to be a physician? What did it mean to be an apothecary? A barber? A surgeon? A barber-surgeon? What did it mean to be licensed to practise medicine? Some of the change had to do with the profession itself: what knowledge or experience was required of or expected from someone to justify the title "physician", "apothecary", "barber-surgeon"? Some of that change had to do with the project of commodifying medication. A medication marketplace does not just require medication as a saleable object. It also requires people to buy it and sell it. Both groups have to be created, but to maximize profit the former has to be infinite and the latter has to be very small. Titles were vital for this project, and no title was more important than "prescription".

Mary Trye and Jane Barker show how and why the prescription system developed into a powerful economic engine. So, let us start with Mary Trye, medicatrix. Biographical facts about Trye are few and far between. She was born in 1642, lost both parents in 1665 during the Great Plague of London, married twice, gave birth to at least one son, lived for a few years in Warwickshire, and returned to London in 1674. Her father, Thomas O'Dowde, trained her to make medication and she became his assistant sometime before she was eighteen. O'Dowde was a fervent proponent of the Scientific Revolution and of chemical medication in particular. He introduced his daughter to the New Science, teaching her to value experience and experiment as well as study. He also trained her in one of the Scientific Revolution's discoveries, chemical medication.

The reason why anyone knows about Mary Trye is because in 1675 she published a small book entitled *Medicatrix, Or The Woman-Physician*. In *Medicatrix*, she claims that she is writing to defend her father as well as the validity of chemical medicine from the aspersions of Henry Stubbe, a Fellow of the RCP. The two parts of the book map out her strategy: rebut Stubbe's argument and extol her father's skills and knowledge in part one, expose the problems in Stubbe's brand of medicine and promote the virtues of her father's brand of medicine in part two. Henry Stubbe is an example of the opposition that the New Science faced. The standard story of the Scientific Revolution has physicians seizing upon the new knowledge and methods, with which they modernize the practice of medicine, create effective treatments, and kick those ugly, warty, cackling crones to the kerb. In reality, some physicians, apothecaries, barbers, barber-surgeons, midwives, and other "irregular practitioners" seized upon the Scientific Revolution, and some did not. The members of the medical community who became science's revolutionaries often became Fellows of the Royal Society; at the very least, they read the Society's journal, *Philosophical Transactions*. Not all Fellows of the Royal Society belonged to the Worshipful Society of Apothecaries or the RCP of course, and most definitely not all guildsmen of the Apothecaries or Fellows of the College approved of the Royal Society: vide Henry Stubbe, Fellow of the RCP, opponent of the Royal Society, and ardent advocate of phlebotomy and studying the ancients like Aristotle and Galen. For someone like Dr Stubbe, someone like Thomas O'Dowde – or much, much worse, Mary O'Dowde Trye – would be quite threatening.

Stubbe was an ideal straw man. He had attacked O'Dowde publicly and had a reputation for vociferous opposition to the New Science. He was also a relatively safe target for a woman with the audacity to publicly attack a man. In 1674, he was drinking heavily and not well respected. Furthermore, although he had published plenty of attacks in the past, he had not published any recently; Trye was provoked by "some Papers [that] came to my hands, subscribed by Henry Stubbe Physician at Warwick". Private papers such as Stubbe's "private Notes and Manuscripts I have by me", even those circulating among a group of interested readers, would not have required a public refutation. Furthermore, the timing is suggestive: "upon my coming to London in October last [1674], being inquisitive after the advance of Chymystry, so desirable by all sorts of

People, some Papers came to my hands". So, after living in Warwickshire for a few years, Mary Trye returned to London and investigated whether the environment was favourable for a woman to establish a practice in chemical medicine. In an answer to her enquiries, someone gave her Stubbe's papers. Perhaps that someone gave them to her with the advice to use them to write against Stubbe and thereby establish herself; perhaps someone gave them to her and she saw the opportunity on her own. Regardless, the papers – private, unpublished – were her opportunity to get her name out to the public and to display her credentials as a medicatrix.

When it comes down to it, however, *Medicatrix* is not about O'Dowde, Henry Stubbe, the Royal Society, or chemical medicine. The book is about herself; she is the eponymous medicatrix, the woman physician. People with a shop put up a sign to indicate where to find them and what they did ("at the sign of the wig", "at the sign of the wheel", and so forth). Trye's book was her shop, and the sign over the door was the word "medicatrix". The term comes from a Latin phrase *Natura corroborata est omnium morborum medicatrix*, which roughly translates as "Nature reinforced is the best treatment for all diseases." The passage appears with slight variations in medical writing throughout the seventeenth century, most often in the final decades. It is attributed to either Hippocrates or Jan Baptiste van Helmont, the most significant contributor to chemical medication after Paracelsus. The word "medicatrix" displays her erudition, but Trye uses it to signal that she is part of the Scientific Revolution, that she uses chemical medication, and that her medications grant her the ability to strengthen nature's curative power. Trye was the first person to use the word as a title for a human practitioner; she invented and conferred it on herself to explain who and what she was. From its title page, *Medicatrix* is not only a declaration of principles, which is bold enough, but also a confident assertion of a self-fashioned identity.

Trye establishes her credentials in ways that her readers would have recognized and approved of. She claims the stereotypical female qualities that help her seem authoritative and rejects those that do not. She reminds readers from time to time that she is a loving daughter behaving with appropriate filial devotion when she defends her father. She anticipates charges that her tract is just stereotypical female chatter (according to one adage, "A woman's tongue wags like a lamb's tail") by promising "to avoid Prolixity, which is a crime we Women are commonly guilty of". Repeatedly, she performs modesty by protesting her inability to judge or make sense of something. One can almost imagine a demure murmur and slyly downcast eyes when she writes, "Now whether Mr. Stubbe be not as conceited in this particular as ever Cicero was, I must leave to those that are proper to judge." Trye also displays the virtues of thoroughness and rigour. "I have given the Reader, this Historical account, not only that he see I have read in History," she explains, "but that I may mind him of the vast difference, between wit and wisdom, truth and errour, justice and interest."

In fact, she delves deep into Stubbe's past, analysing his actions and how he characterizes himself. She uses Cicero's interrogation techniques to challenge Stubbe's self-representation as a well-known, prolific author. "Where hath Mr. Stubbe lived all this while?" she asks. "At Jamaica; or where are his famous works extant, and victorious Books exposed to Sale? For I am inform'd, the Author himself, with most Book-sellers in this City, is not known; and the Books themselves scarce with any to be had; so that I am satisfied, the Generality of this

Kingdom never heard his name, much less saw him." She shows off her own knowledge of chemical medication and her connection to men, such as in her alliance with the Royal Society.

Trye's characterizations of O'Dowde and Stubbe establish her own integrity, dedication, skill, and courage. When Stubbe and other physicians fled the plague, she and her father remained. Consequently, Stubbe saved no one, her father saved vast numbers of the public, and Mary saved members of the public and her own family. Her fearlessness appears also in her attacks on Stubbe and his like-minded colleagues, calling him names (for example *medicus*, "the Great Oracle Mr Stubbe", "the Tinkling Campanelle", "the Quacking Parrot"), mocking his accomplishments ("O prodigious!" she snorts at one point), and accusing him of being more talk than substance. "Learning will fit a Man for that Profession, but a diligent and indefatigable Elaboration must perfect it, Medicines when obtained, one may in a reasonable time learn to apply," she writes, "but how to obtain those Medicines, I verily think is a question beyond Dr. Stubbes's Study." She, however, has the combination of education and experience vital for serving the ill:

Yet I do say, Learning in it self, is only preparatory, not perfect, a proper progress and tendency, in order to the Art of Physick, not the Perfection and Consummation of that Art: A Man may read and Author, and yet not understand a medicine [...] And as I am not satisfied, That every Author that writes of Medicines understands them; so I am as well assured, That a Man may sleep many years at the Fountain of Learning, and yet awake no Physician: Medicines are the Marrow and full Perfection of a Physician, and those are hard to be attain'd...

These are the qualities of a physician, and as *Medicatrix* makes clear, she has them. She knows what medicinal treatments should be used for what ailment, and equally important, she knows how to make those treatments herself. She is a chemical physician. A medicatrix.

As a declaration of self, Trye's book is stunning. It does not just assert, it displays her equality with physicians and apothecaries. It rejects the accepted ways to fulfil the requirements for a licensed or guild membership: studying at a university (she did the reading on her own), being fluent in Latin (Who needs it? She read excellent translations), apprenticing to a member of a guild (she apprenticed to her father, who was a master chemical physician), gaining experience (she assisted O'Dowde and treated people herself), learning what medicines treat which illnesses (she was apprenticed to O'Dowde), and learning how to make those medicines (again, apprenticed to O'Dowde).

Furthermore, *Medicatrix* is a declaration of competition, and as such quite alarming. Her justifications for writing – I am a good daughter vindicating my wronged father; I am a charitable, knowledgeable person seeking to help people; Henry Stubbe is a cowardly quack who must be slapped down – enable her to promote herself and her medications. "Mr. Medicus hath forc'd me to tell the World, in answer to him, what Remedies I can afford them, and what good I can do them," she complains. Yes, right, of course it is Henry Stubbe's fault that she must tell everyone how effective she and her medications are. The last eight pages of *Medicatrix*, entitled "An Advertisement of Dr. O. Dowdes Medicines, and the Authors", list medicaments that Trye sold for several common, painful afflictions, including gout. Between this section and the end of her treatise, Mary Trye included a two-page postscript, in which she challenges

Stubbe "to an experimental Tryal" to prove that what she has said is true. Like the opening, the language of the postscript tends towards the formal challenge ("I am and shall be ready to maintain" and so forth), another way of making Trye's position more male than female.*

Mary Trye and *Medicatrix* highlight two related problems that accompanied the challenge of commodifying medication. The first was excluding women and their medication from the medication marketplace. The second was shifting the culture so the professional's medication was the one that everyone sought. The prescription system solved both of these problems. Prescriptions had been around for centuries. From the mid-seventeenth century onwards, professionals used them to create an economic system: the physician got money for writing the prescription and the apothecary got money for filling it. Legally, only a physician could write a prescription and only an apothecary could fill one, which should have restricted participation in the system to men. The physicians and apothecaries of London wrangled with each other over the privilege of diagnosing and prescribing, but they united when it came to women's exclusion from the system.

The prescription told the apothecary what medication was required, including the ingredients and their quantities; the dose to be taken, how often, and in what form; and the duration of the medication's use. Until well into the twentieth century, it was

---

* Trye was not the only woman in London challenging someone to a duel. That same year, Hortense Mancini, Duchess of Mazarin, and Anne, Countess of Sussex, engaged in swordplay one night in St James's Park, wearing only their nightgowns. The duchess had left her much older, insane husband in France and so could do as she pleased, but the countess was hauled off to their estate in the country by her husband, who was, in the words of a future monarch, "not amused".

standard for the apothecary (or druggist, chemist, or pharmacist) to make up the medication themselves. The doctor wrote the prescription for a medicinal compound and the apothecary either knew what ingredients were required or followed the instructions in the script. He would then mix it together and dispense the compound to the patient.* In addition to the legal restriction, the prescription system excluded women and other untrained people by virtue of its elaborate code. Medical students learned it in their courses; notebooks reveal that they even took notes in it. Apprentice apothecaries learned it from their masters.

The situation with women and guilds was messier than it was with women and universities, where they were forbidden. Some guilds in some towns at some point admitted girls as apprentices, and some granted women the "liberty" or "freedom" to act as a member if they proved their knowledge and skill. Esther Dudley, a widow, was "admitted into the freedom" of the coopers' guild because her father was a cooper. Some guilds allowed widows to take over the business and some also allowed widows to take apprentices. It was difficult to remain ignorant when the shop was inside the house and the wife of the guildsman was also keeping the books and caring for the apprentices. Nevertheless, women's part in guilds, including the apothecary guilds in London and beyond, was not consistent or sizeable enough for them to be real participants in the prescription system.

Measurements were the first level of the cypher. Dry

---

* This is how Mr Gower nearly poisons Mrs Blaine's little boy in the film *It's a Wonderful Life* (1946). (Fortunately, despite not being trained as a pharmacist, twelve-year-old George Bailey manages to read the word **POISON** written in giant letters on the bottle and prevent disaster.)

ingredients such as powders were measured in grains, indicated with "G" or "gr"; scruples (Э); drachms (ʒ); ounces (ʒ); and pounds (lb). Occasionally, there was a call for "handful" or "pinch", symbolized as "M.j." and "p", respectively. Wet ingredients were measured in minims, indicated with "m"; guttes or drops (gtt); drachms (ʒ); ounces (ʒ); pints (O); quarts (d); and gallons (C). Yes, that's correct: both wet and dry substances were measured in drachms and ounces. Pints were abbreviated "O" because another name for a pint was octuarius, and gallons were signified with "C" because another name for a gallon was congius or "congy", as it was sometimes pronounced. Numbers were written as Roman numerals (i, v, x), except if the i was on the end of the number; then it was written as j. Thus 1 was written i, but 2 was written ij and 6 was vj. Often the "v" after an "i" was written as "ÿ", however, so 4 might be iÿ. To indicate half of an amount, the writer could use "ss" but much more often used "β", which was faster and easier to write (especially with a quill). Half a scruple had its own name: an obol, which mercifully was written "obol". Most of the time, however, half a scruple would be written as Эiβ. There were also measurement shortcuts. "Ana" indicated that all the ingredients in a list should be added in the same quantity; "p.ae." meant that the product should be divided into or administered in equal portions.

Students and apprentices had to learn the symbols and abbreviations for different elements, metals, and transformed metals, as well. The symbols came from alchemy, which was the starting point for the modern discipline of chemistry and for chemical medication. Salt, for example, was written as Θ, sulphur was ♀, and philosopher's sulphur was ♠. Other common symbols included mercury (☿), silver (△), and antimony (♁). Latin supplied the vocabulary for a medication's effect, such as astringent (closing a wound), pectoral (an expectorant), or sudorific (provokes sweating), although the ordinary housewife had a chance of recognizing words like these even if she did not use them herself. When it came to the forms that medication took, however, there was a shared lexicon. Medical terms such as clyster (enema) or linctus (medication to be licked off a spoon) were also household terms found in recipe books as well as in physicians' and apothecaries' records. Put together, a prescription often looked like gibberish. Take this one, written by an apothecary in 1740:

Sal. Vol. Oleos ℨss Spt. Lavend. c. ℨi
Tinct. Castor. ℨij. Misic.
xxv Drops to be taken Morning and
Evening in a Glass of white wine and
water;   ad. Vertigineum

Who could understand this? Certainly not the patient, their family, neighbours, or friends. Hopefully their vertigo went away, at least.

The *Pharmacopoeia Londinensis* was the last layer of code. Although one apothecary, Nicholas Culpeper, went rogue in 1650 and translated it into English so housewives could use it, the official editions remained in code, and for some time, in Latin. Furthermore, the *Pharmacopoeia* did not indicate quantities or processes. The apothecary who was instructed via prescription to make Æthiops Mineralis would read, "Rx Mercurii crudi, Florum Sulphuris, ana partes aequales. Agitentur in Mortario vitreo, cum pistillo virtreo, donec Mercurius evanescat": mix equal amounts of mercury and flowers of sulphur in a glass mortar with a glass pestle until the mercury is fully integrated.

But how much mercury? How much flowers of sulphur? The apothecary often knew offhand; some compounds were so common and made so frequently that it would become almost automatic. Knowing this, some physicians wrote "Pharm Lond" on their prescriptions to tell the apothecary to follow the instructions in the *Pharmacopoeia*. Otherwise, the prescription would tell the apothecary exactly how the physician wanted it done.

There was a significant potential for error in this system. Doctors have always been famous for bad handwriting, so reading a prescription was difficult enough without having to remember the meanings of the abbreviations and symbols. After all, quantity mattered a great deal with ingredients. Some were fairly benign no matter how much was ingested, but others such as poppy or foxglove could be fatal if taken in the wrong amounts, and killing one's patients was definitely contraindicated. The potential for disaster was worth it. The code enabled the professionals to indicate that their medicines were fundamentally different from those made by women, and to hide when their medicines actually were not.

At the start of the Scientific Revolution, most medication, regardless of who made it, came from organic materials, primarily plants. If the primary ingredients for making medicine could be grown in anyone's back garden or picked up along the side of any local road or river, and if they were often prepared the same way to make the same treatment, then there was not much difference between the ingredients of an apothecary's medicine and the ingredients of a housewife's medicine. Not much or none. That was not something the professionals wanted anyone to know, and the code hid it. As historian Patrick Wallis puts it:

Knowledge of many such substances was widely diffused, and formed part of the skills expected of women in particular. Yet in the apothecaries' and druggists' hands, these commonplace materials somehow also became expensive and exclusive, a transformation of muck into brass that seemed to many a uniquely, and unjustifiably, profitable enterprise.

"Capiscum" sounds so much more impressive than "pepper". So does "Borrago floribus caerulis et albis" for blue and white borage flowers. True, apothecaries occasionally used exotic ingredients that ordinary women would not be able to find or afford. In addition to "Powders of Viper" and snake's skin, a list of unusual materials compiled by Robert Pitt in 1703 included "Bezoar-stone", pearl, gold, silver, gemstones, mummy (ground-up bits of mummies – yes, really), pike's jawbone, swallow's nest, "the Bone in the Heart of a Stag", rabbit and boar ankle bones, elk and "ounce" hooves, and "the Horns of the Elke, Buffalo, Rhinoceros, Unicorn". Pitt was not an anomaly. John Quincy also included unicorn's horn and "bone of a stag's heart" in *Pharmacopoeia Officinalis & Extemporanea* (1724). Imperial expansion ensured that there was always a new, hard-to-get substance from a distant corner of the earth that might be said to have medicinal properties. "These are some frontier plants as it were, that cannot with any certainty be ranked amongst ither [*sic*]," one medical student recorded in his lecture notes, but they were still included in the lecture on medicinal botanics. Tobacco, coffee, the "Magellanic Bay-like tree", the "India Berry tree", and the euphoniously named "Gamboodge tree", for example, were used to cure everything from cancer to leprosy to lice to sexually transmitted infections.

Unfortunately for the professionals, imperial expansion also meant that it got easier and easier for ordinary people to acquire those exotic substances (except perhaps unicorn horn). Quincy noted in *Pharmacopoeia Officinalis & Extemporanea* that although Pliny thought red coral was "found only in the Indian Seas", in fact "we have it now from many parts of the Mediterranean, and Naples is a great Market for it." By the end of the seventeenth century, a woman's recipe might call for cinnamon, nutmeg, or ginger whether she was baking a cake or making a bolus. The code, however, preserved the mystery of a prescription's ingredients. What's in a name? Profit, my girl. Profit.

Then there was "human skull". In 1724, John Quincy explained disdainfully in *The Complete English Dispensary* that it "has obtain'd a Place in Medicine", although "more from a whimsical Philosophy, than any other account". The philosophy might have been whimsical, but many professionals and housewives accepted it. In *The Marrow of Physicke* (1648), Thomas Brugis recommended "Oyle of Mans Skull" for epilepsy, instructing his readers to "buy this Oyle of the Chemists". If a chemist was not available but a skull was, unlikely as that seems, Brugis also had a recipe:

> The Skull of a man that hath been dead but one yeare, and bury it in the Ashes behinde the fire, and let it burne untill it be marvellous white, and so well burned that you may breake it with your finger; then take off all the uppermost part of the Head to the top of the Crown, and beat it as small as is possible, then grate a Nutmeg, and put to it, then take Dogs blood, and dry it, and make Powder thereof, and mingle as much with the other Powder, as the Powder weighes, and give it the sick to drinke, both when

he is well, and when he is sicke, first, and last, and it will help him by Gods grace.

As in professionals' books, "skull" turns up rarely in recipe books, but every now and then a recipe calls for the skull of a man – sometimes specified as "a dead man", as if one could get a man's skull under other circumstances. To make lozenges, one recipe required coral, pearl, amber, crab eyes, crab claws, hartshorn, ivory, mistletoe, and "Man's skull". Hartshorn and mistletoe were rare but not bizarre ingredients; everything else, especially the skull (hopefully), was. Recipes did not always require the whole skull, fortunately. Another woman's recipe, this one for "Convulsion and fitts" of a pregnant woman, begins: "Take one ear off dead Mans Skull yt was never buryed." Frankly, it is unlikely that anyone at home actually made any of the recipes requiring pieces of a dead man's head. A woman need not have believed in its curative properties to collect the recipe, however; recipes involving "dead Mans Skull" had a titillating "eew" factor even then.

The fact is, a prescription was really a recipe, a truth that the professionals did not want their customers to realize. Part of their project was to replace "recipe" with "prescription" in the general lexicon. For centuries, professional makers of medication used the word "recipe" far more often than "prescription". The word "recipe" was substantially more common in printed medical materials than "prescription" until the last few decades of the seventeenth century, when they began to switch places. In 1701, for example, an anonymous author of *The Present State of Physic* extolled physicians' "prescriptions"; George I used the word in his 1721 proclamation on medicine.

The transition was not just about swapping one male-associated word for another female-associated word. By the

1730s, "recipe" denoted the list of ingredients and procedure, while "prescription" referred to the document presented to an apothecary or to the product. *The Generous Physician*, printed in 1733 and attributed to Sir John Colbatch, Fellow of the RCP, offered "The Best Receipts in English, and Directions how to use them, adapted to ordinary Capacities" on its title page, but the preface concludes with the assurance that "No Prescription is embarrass'd, or Ingredient invented, to bring on an Application to a Physician." In Daniel Bellamy's comedy *The Merry Swain* (1739), a sighing lover asks a cynical friend for his "Receipt" to cure love, and on receiving it exclaims, "I like your Prescription." The shift to "prescription" was a shift in focus to the product and away from the process indicated by "recipe". By 1740, *A New General Dictionary* defined "prescription" as "the appropriating proper remedies to particular diseases goes by this name; also the medicine itself".

Built into the concept of a recipe was the idea of communal knowledge. Recipes were supposed to be shared with others. In contrast, prescriptions were built on the idea of knowledge as property, so they were extremely limited in circulation. To the great delight of the RCP, in 1721 George I issued a proclamation "Commanding Apothecaries to follow the Dispensatory lately compiled by the College of Physicians of London", although he permitted some variation: "except it shall be by the Special Direction or Prescription of some Learned Physician". Mary Trye subscribed as much as any proponent of the Scientific Revolution to this view of knowledge and medication. She touted her own cures in the "Advertisement" section of *Medicatrix*, but she did not say how they were made. After extolling her training, knowledge, and skill, the book ends with a catalogue of goods. The whole thing is an advertisement, not just the last eight pages. She had to promote her ready-made medications, however, because as a woman she could not write prescriptions.

One of the diseases that Mary Trye discusses in *Medicatrix* is gout. Gout is a highly painful condition caused by uric acid crystals building up in the joints, most often those of the big toe. It feels something like a herd of sea urchins wedged between two bones and set on fire. James Gillray painted its portrait (see plate section, p.4). Gout was a "disease of the rich" because it was caused by a fatty, sugary diet and sedentary lifestyle. It was also considered a man's disease because women's diets contained less alcohol (and because menstruation was thought to purge women of some of the toxins that caused gout). When Queen Anne was diagnosed with gout as a young woman, probably for lack of a better explanation for her ailments, the diagnosis implied that there was something unnatural and unwomanly about her, an unsurprising accusation of any female monarch, but especially pointed given her inability to provide an heir to the throne.

There was quite a demand for gout treatments because none of them seemed to work. In his book *The Attila of the Gout* (1713), the surgeon John Marten called it "the great uncertainty in which Physicians fluctuate as to the true Method of Cure", or as Jane Barker wrote, "The sturdy *Gout*, which all *Male* power withstands". A seventeenth-century adage put it more astringently: "The Physician is blind at the Gout." Hermannus Vander Heyden advocated cold water, Roger Dixon used "Sol Triumphans, or Horizontall Gold" (whatever that is), and Benjamin Welles recommended purges and phlebotomy.

Handbooks of medication listed multiple substances or recipes that could ease or cure gout. The *Pharmacopoeia Londinensis* of 1702 suggested among other materials powdered bear's breech, the roasted roots of Hounds Tongue, oil of ebony, guaiacum, germander, the fat or marrow of a dog, the gall of a puppy mixed with vinegar, and the flesh or excrement of a cat. Many of the *Pharmacopoeia*'s ingredients contribute to the gout medications in Fuller's *Pharmacopoeia Extemporanea* (1714). His "Arthritic ale" included guaiacum and germander, as well as avens roots, agrimony, sage, betony, raisins, sassafras, ground pine, "Hermodactyls", "Dodder of Thyme", "Stechas Flowers", and "Wort". For gout, William Salmon recommended alabastrum unguentum, made with red briar, white wine, rue, chamomile, powdered alabaster, fennel seeds, oil of roses, wax, and five egg whites. Mary Trye sold "A Medicinal Milk, an Aural Tincture, Two sorts of Radiant Pills. A Purging Potion, an Extract against the Gout, A Cordial Potion; and for outward application, Two Unguents, the one White, the other Green, and my Golden Balsamick Spirits", all of which removed gout pain "in Three or Four days".

Treatments for gout appear in women's recipe books, although they are considerably rarer than medications for less class-dependent afflictions, such as cough, burns, cramps, headache, eye trouble, and labour pains. Some gout treatments made a more credible attempt than others: one called for harvesting the root of a male peony during a waxing moon in May to wear at one's waist. One type of gout medication was meant to be ingested, a convenient delivery method for an ailment that made the afflicted place unbearably sensitive. 'Mary Courthop's recipe for a gout drink mixed "venice [*sic*] Turpintine" with egg and milk, for example, while Lady Sedley's cocktail combined turpentine, wax, wine, rose water, and salad oil. "Turpentine" was then what it is now: a thick, resinous oil made from the terebinth, a kind of tree. These days it is used for strengthening horse's hooves and creating a shiny coating over paint.

There were also topical treatments for gout because like burns and acne, it involved the skin and flesh, but anything that had to be smeared on, such as an ointment or a salve, was of very limited use because the place was so sensitive. A more comfortably applied treatment was a plaister, a very light cloth that had been soaked in a compound of melted fat or wax and medicinal substances such as herbs, and then dried. The cloth (also called searcloth, serecloth, or cerecloth) would be laid over the swollen area; its healing properties would sink into the skin as body heat melted the wax and released the herbal compound. According to a late seventeenth-century recipe book compiled by Mary Granville and her daughter, Anne Granville D'Ewes, "Mrs Badge's Plaister" required both "white Rosin" (resin) and "the best yellow bees wax" as a solution for the rest of the ingredients. Plaisters could be made in bulk by preparing a long length of cloth from which pieces could be cut as needed (as Lettice Pudsey advised, "role up the seare cloth: & keep it for your use").

Such gout treatments were often part of a multi-cure medication. Lady Sedley's powder, to be consumed by the spoonful for "three dayes following fasting", was good "For the Dropsy in the Womb, or in the feet, or for the Imposthume in the Stomach or Gout in the Stomach, or for any evill". Lady Barkham's "oyntmt" was good "for aches: gout, pyles, swellings, bruses: spraynes, the Kings Evill: and divers other grevances". Lady Gifford's cerecloth treated "gowt" as well as "aches: [and] greene wounds", and could "strengthen broaken bones", while another recipe for a "serecloth" was to be used "for any ach in any place: broken bones: for the gout proved: it doth seldom fayle to ease the gout". As medicine and chemistry evolved according to the Scientific Revolution's principles, the idea of a panacea – a medicinal philosopher's stone – became, also like the philosopher's stone, increasingly fantastical. Perhaps desperation suggested that a medication that did other things could also cure gout, or vice versa. Perhaps a cure for gout retained the mythical status of the philosopher's stone because it proved so elusive.

In 1685, bookseller Benjamin Crayle advertised in the back of one of his books for sale that "Dr. Barker's Famous Gout Plaister" could be purchased at his shop in Fleet Street (look for the sign of the lamb). The back pages of a volume of titillating narratives seems like an odd place to hawk medication, but it was common practice to use that part of a book as an advertising section. That is one reason why Mary Trye listed her own medications for sale at the end of *Medicatrix*. Crayle also advertised Monsieur Blegny's "Venereal Water" in the back of *England's Heroical Epistles* by Michael Drayton, along with a long list of books, some of them on the same topic, such as *Philaxa Medicina*, *The Queens Closet Opened* (which I discussed in Chapter 2), and *New and Curious Observations on the Art of Curing Venereal Disease* by Monsieur Blegny. *The Queens Closet Opened* was also one of the books advertised in *Delightful and Ingenious Novells* with "Dr Barker's Famous Gout Plaister".

For five shillings per roll, the Famous Gout Plaister "infallibly takes away the pain in Twelve Hours time, with the Paroxysm of the Distempter, and in time may effect a perfect Cure". Twelve hours might not sound particularly efficacious, but relief in any amount of time would have been appealing. Besides, some remedies claimed to take even longer. "The Duke of Portland's gout powder" took two years. Similarly, the high price for Barker's "gout plaister" is not so very exciting. Five shillings was five times the amount usually charged for advertised medication, equivalent to roughly two days of wages for a labourer. However, labourers did not get gout: rich people did, and rich people could and would pay a premium to ease their agony. Besides, a whole roll of plaister provided multiple doses.

What is outrageous about "Dr Barkers Famous Gout Plaister" is that Dr Barker's first name was Jane. Then again, Jane Barker was an outrageous person. As a teenager, she sent a perfidious ex-suitor a pair of horns on his wedding day, "in the Presence of the Bride and all the Company; as also several Emblems and Mottos on the Subject, the Horns being fasten'd on a Head-band, as a sovereign remedy for the Head-ach, to which marry'd Men are often very subject, especially those who marry Town-Coquets".

It was not surprising that she knew a great deal about medication. She grew up in the country and had the conventional upbringing of a landed gentleman's daughter, as recipes for punch, flummery (a sweet dessert, often in the jelly family), and chicken soup in her novel, *A Patch-Work Screen for the Ladies* (1723), attest. She learned reading, writing, mathematics, and the requisite domestic skills and knowledge for marriage within her class, including how to treat the ailments and injuries of a family and its dependents. She was taught about medicinal plants and worked in the kitchen, learning to transform those ingredients into salves, pastes, cordials, boluses, tablets, electuaries, fomentations, and glisters. When a medication did not work, she and her mother would have asked friends, neighbours, and tenants for other recipes, and would have experimented with their own to find one that did. As a girl, Barker modelled herself on Clorin, the title character in a popular play, *The Faithful Shepherdess*, who healed the other characters of illness, injury, and heartbreak.

Barker knew much more than domestic medication, however. She was also up to date with the Scientific Revolution's methods and discoveries, thanks primarily to her own fierce determination and secondarily to her older brother, Edward, who studied medicine at Oxford, Leiden, and Paris. Initially when Barker asked him to share his knowledge, Edward refused, claiming that he could not teach anyone who did not know Latin. So she set about learning Latin – from Edward: "I got my Brother, who was not yet return'd to *Oxford*, to set me in the Way to learn my Grammar, which he really did." However, "thinking it only a Vapour of Fancy" that she wanted to study medicine, he expected her to give up when the Latin got difficult. He underestimated her. She stuck with it, and Edward Barker kept his word: he taught her what he had learned whenever he came home. As she described it in her semi-autobiographical narrative *A Patch-Work Screen for the Ladies*, "[M]y brother continued to oblige my fancy," and "assisted me in Anatomy and Simpling, in which we took many a pleasing Walk, and gather'd many Patterns of different Plants, in order to make a large natural Herbal". In a poem to him on his medical studies in France, Barker lamented that in his absence:

> Nothing at present wonted pleasure yields,
> The *Birds* nor *Bushes*, or the gaudy *Fields*;
> Nor *Osier* holts,* nor Flow're banks of *Glen*;
> Nor the soft *Meadow-grass* seem *Plush*, as when
> We us'd to walk together kindly here,
> And think each blade of Corn a Gem did bear.

Under Edward's tutelage, she obtained the latest medical knowledge. "I made such progress in Anatomy, as to understand Harvey's Circulation of the Blood, and Lower's Motion of

---

* An "Osier" is a kind of willow tree.

the Heart," she wrote proudly, explaining that "My Time and Thoughts were taken up in Harvey, Willis, and such-like authors."

These authors were advanced knowledge for anyone, and certainly for a woman at the end of the seventeenth century, but she had Edward and she had Latin: "my Brother help'd me to understand and relish [them], which otherwise might have seemed harsh or insipid". William Harvey's *Exercitatio anatomica de motu cordis et sanguinis in animalibus* (1628) revealed that blood circulated through the body, a tremendous breakthrough. It had been translated from Latin into English in 1653, but she was probably reading Edward's copy and he probably had it in Latin. "Willis" refers to Thomas Willis, who was a founding member of the Royal Society and authored *Cerebri anatome, cui accessit nervorum descriptio et usus studio Thomæ Willis* (1664) and *Pathologiæ cerebri, et nervosi generis specimen in quo agitur de morbis convulsivis, et de scorbuto* (1667), on brain anatomy. She must have read those and the 220 pages of Richard Lower's *Tractatus de Corde* (1669) in Latin because none was translated into English until the 1680s.

Edward Barker died unexpectedly after a short, intense fever. He was twenty-five. Jane was twenty-three, and devastated. Eulogizing him, she called Edward "him [who] my Soul ador'd with so much pride, / As makes me slight all worldly things" in grief. He left behind a personal library, papers, and notebooks from his medical studies, materials that no one in the family wanted or cared about except Jane. Initially, she turned to his books and papers for solace, although "No Book or Paper cou'd I turn over, but I found Memorandums of his Wisdom and Learning." Over time, her interest in studying medicine revived and she used everything that Edward had amassed over the course of his medical education – "those Books, on which I had seen my Brother most intent" – to complete hers.

Barker was very aware that this background made her a hybrid of new and domestic medicine, and her writing celebrates her multifaceted body of knowledge and practice. Although her most recent biographer, Kathryn King, calls Barker's view a "confused, shifting, and contradictory play of scientific under-standings", she was not at all confused. She was forging a radical union of the New Science and the old, domestic, female-associated model of medicine. Consider Barker's description of making a herbal with Edward: "we took many a pleasing Walk, and gather'd many Patterns of different Plants, in order to make a large natural Herbal". Women would have needed to know medicinal plants, especially those that grew in the vicin-ity, and walking around the countryside gathering plants was part of a woman's tasks. Those walks might have been pleasant – one hopes so – but taking "many a pleasing Walk" to gather "many Patterns of different Plants" is an undomestic way of representing them. Herbals themselves were increasingly the tools of science's revolutionaries as they attempted to list and sort everything in nature. In Barker's hands, female knowledge and skills are equal partners with male knowledge and skills.

Barker's overlap of male-associated New Science and female-associated domestic medicine is most pronounced when she discusses medication directly. Putting anatomy and "sim-pling" together as the two subjects she studied with Edward, for example, pairs the Scientific Revolution's discoveries with traditional medicine's methods. A "simple" was a medicinal liquid made from a single organic ingredient. Simples and simpling were foundational to both domestic medicine and professional medication; much of early medical chemistry went into developing simples from inorganic materials such as mercury and sulphur. In her poem "On my Mother and Lady W. who both lay sick at the same time under the Hands of Dr. Paman", Barker contends that Dr Paman's "great *Art*" lies in his medications, his "best Receipts", the part of physicians' practice that doctors shared with women. Barker's fervent wish for Dr Paman is that no apothecary should mess with his recipes: "Nor to bad Druggs let Fate thy Worth expose, / For best Receipts are baffl'd oft by those."

Confident in her knowledge and skills and unsurprisingly still unmarried, Barker took her talents to London. Big-city newspapers like the *London Gazette* and smaller periodicals such as the *Ipswich Journal* and the *Newcastle Courant* were full of advertisements for medication and remedies such as *Eradicatorium Arthiriditis*, Pectoral Drops, or the enticingly labelled Royal Purging Cordial. Asthma, headache, rheumatism, deafness, infertility – newspapers boasted cures for them all. Barker was aware of the dangers of quacks, people selling substances with purported rather than actual curative powers. As she wrote in her poem "On my Mother and Lady W. who both lay sick at the same time under the Hands of Dr. Paman":

> Nor let no Quack intrude where thou do'st come,
> To crop thy Fame, or haste thy Patients doom;
> Base Quackery to Sickness the kind Nurse,
> The Patients ruine, and Physicians curse…

She was no quack, however. Domestic medicine's recipes were time-tested; as far as she was concerned, there was nothing speculative or fraudulent about them. Furthermore, Barker also had an up-to-date medical education, thanks to her brother Edward. So off she went, and by 1685 she was selling gout medication under the name of Dr Barker.

There was no need for a woman advertising her medication to hide behind a title, false or otherwise. Plenty of women used their own names when advertising their medical expertise or treatments. A full name was not even necessary; "Barker's Famous Gout Plaister" would have sold just fine. For Barker as for Mary Trye or any man, the title was a statement of identity, of credentials. After all, what entitled a man to call himself "Doctor"? She had read what her brother had read, she knew what her brother had known, she cured people's ailments like her brother had, and Edward had been a doctor. Ergo, she was a doctor. But Barker went further. She prescribed medications to be made by others. Even better: those others then filled her prescriptions. And sometime in the early 1680s, probably not far from St Paul's Cathedral in London, she recorded this triumph in a gleeful poem entitled "On the Apothecary's Filing My Bills Amongst the Doctors". Jane Barker had written a prescription and an apothecary had taken it for the real thing, a prescription written by a licensed male physician. It was bad enough for the nascent prescription system when apothecaries learned prescriptions for different illnesses through filling physicians' orders. As Gallypot announces in *Physick Lies A-Bleeding*, "I can write a Prescription as well as any of 'um all, I learn'd that the first thing, by reading Doctors Bills in my Shop." Jane Barker was not an apothecary, however; she was a woman, which was so much worse, and such behaviour was an outrage. It would have been had anyone found out, anyway.

Barker did not consider herself *equivalent* to a physician. As she wrote in "On the Apothecary's", she had joined their ranks – she *was* a doctor:

I hope I shan't be blamed if I am proud,
That I'm admitted 'mongst this Learned Crowd;

125

To be proud of a Fortune so sublime,
Methinks is rather Duty, than a Crime...

It is an important distinction. As she sees it, she has not fooled anyone into thinking that she is both a physician and male. She has been "admitted" to the "Learned Crowd", who themselves "exalt and own our Fame", and "gain'd this mighty place / Amongst th' immortal Æsculapian Race". As she puts it succinctly about the apothecary's action, "This tis, makes me a famed Physician grow." Nor is she accepting her triumph with demure decorum. "But with this honour I'm so satisfy'd, / The Antients were not more when Deify'd," she crows, calling her admission to the company of physicians (the "Aesculapian Race") "a Glory that exceeds excess" and "transcends all common happiness".

In addition to doing something that the men could do – that is, write a prescription – Barker did something that they could *not* do: prescribe a gout medication that worked. "The sturdy Gout, which all Male power withstands," she explains, "is overcome by my soft Female hands." (Mary Trye made a similar claim in 1675.) Considering the number of people and how badly they suffered, Barker saw her success in biblical terms:

Not Deb'ra, Judith, or Semiramis / Could boast of Conquests half so great as this; / More than they slew, I save in this Disease.

She also compared her joining the ranks of physicians to "Saul [who] 'mongst Prophets turn'd a Prophet too". It was not enough to be as knowledgeable and skilled as a physician if one were also female. She complains that "Some Women-haters may be so uncivil, / To say the Devil's cast out by the Devil" when a woman's medication worked, but patients felt differently: "so the good are pleas'd, no matter for the evil". If the medicine works, what does it matter who made it?

It matters a great deal if the emphasis is on the medication rather than on the healing. Mary Trye referred to Henry Stubbe's "dis-ingenious and inhumane Brethren, that care not what becomes of Sick [*sic*], or any thing else, so they can support their own Grandieur, Profit, and Interest". Jane Barker recognized that the exclusion of women from the prescription system had everything to do with money. As she cynically explained in "On the Apothecary's", "Thus Gold, which by th' Sun's influence do's grow, / Do's that i'th' Market Phoebus cannot doe" ("Thus Gold, which by the Sun's influence does grow / Does in that Market what Phoebus cannot do"). The rising sun resembled a gold coin, but it did not have money's power in the drug market. It was a not-so-subtle jab at the very uncelestial motives of the professionals. Phoebus Apollo was the Greek name for Apollo in his charioteer-of-the-sun aspect, and Apollo was also the God of Healing. The God of Medicine, Asclepius – or as Barker spells it, "Aesculapius" – was his son. She uses the same terms in other poems, such as when she calls her late brother's colleagues "the Apollo's of thy noble Art" and that "Gallant Aesculapian Crew". Describing her early ambitions to study medicine, she wrote, "Thus I sought to become Apollo's darling Daughter." In "On the Apothecary's", Barker unsubtly asserts that neither Phoebus Apollo nor anything divine had a whit to do with "That Market". Anyone who cares about healing will agree with her that "so the good are pleas'd" is all that matters, but the priority for professional healers is money.

The prescription system was not established overnight, of course. It took time, for example, for "prescription" to replace "recipe" in the popular imagination and vocabulary. Barker, it will surprise nobody to learn, eschewed the word, instead celebrating Dr Paman's "Receipts" and her own ("On the Apothecary's"). It also took time for the professionals' mystery medication to seem as good as, if not better than, women's domestic medicine. "How oft, when eminent physicians fail, / Do good old womens [sic] remedies prevail?" asked James Bramston in 1733. "Of Graduates I dislike the learned rout, / And chuse a female Doctor for the gout." Not everyone accepted exclusion and obfuscation as the foundation for a system of medication, either. "Sophia, a Person of Quality" argued that "Reason is absolutely unlimited in her jurisdiction over mankind; we are all born to judge of what concerns and affects us, and if some cannot use the objects of sense with the same facility as others, all have an equal right to them." Knowledge, including the New Science and prescription codes, cannot be "diminish'd by communication". Everyone benefits when everyone contributes to the search for "Truth and knowledge"; those attempting to keep women away from the sciences are interested only in themselves.

There were practical ways of resisting. A raft of books began to appear in the late seventeenth century that explained the meaning of apothecaries' symbols, translated medical terms, and helped people buy medication or ingredients for medication from apothecaries. William Salmon, a prolific producer of medical and housekeeping manuals, included a dictionary of medical terms and a list of tools for making medication at home in the back of his *Family-Dictionary* (1696). The second edition of John Quincy's *Lexicon physico-medicum: or, a new medicinal dictionary; explaining the difficult terms used in the several branches of the profession, and in such Parts of Natural Philosophy as are introductory thereto: with an account of the Things Signified by such Terms* (1722) included a chart of apothecaries' symbols. Pages like these disappeared from these kinds of books, however, not only because the opponents of secrecy faded away but also because women were increasingly discouraged from dealing with illness at home (see Chapter 2). The separation of medicine from food, women, and the kitchen is reflected in and effected by printed materials, from the amount of space allotted to and the kind of medicinal recipes in housekeeping manuals to the inclusion of lexicons and charts to the definitions of words.

Trye and Barker were part of the rearguard action being fought for women's traditional position. In promoting their own expertise, they emphasized that their knowledge came from and with the approval of men. Trye justified publishing *Medicatrix* by claiming filial virtue: she owed it to her father, good girl that she was (although she called herself his "only child" rather than his daughter). Barker gloated in verse, but it was unpublished verse. Her published works – the semi-autobiographical narrative trilogy – emphasized her brother's role. The two women exemplify the union of what the professionals were trying to divide into separate and very unequal spheres. They also illustrate the variety encompassed by the term "Scientific Revolution". Trye was married with children, part of the Scientific Revolution, and skilled in chemical medicine. Barker was single without children, part of the Scientific Revolution, and skilled in organic medicine.

Readers know how this story ends: the prescription system became the method for getting medications with the best reputation, elevating prescription medication over homemade or store-bought. The system thickened the cloud of mystery around professionals' medications and made the unknowability of medicine's components one of its positive features. Jane

Barker's triumph "upon the apothecary's filing her bills among the doctors" demonstrates just how difficult it was to enter into the prescription system's closed course as early as the 1680s. She is indeed the exception that proves the rule.

Trye and Barker had one more thing in common. Both worried about what impact commodifying medication would have on people who could not afford it. Barker praised Dr Paman for being "mightily approv'd, / Both by thy Patients, and the Poor belov'd". Excluding women from the prescription system was bad enough; persuading people that women's treatments were useless, dangerous, or fatal not only destroyed the viability of a set of medications but also helped to destroy the system for distributing them. Without domestic medicine, what would happen to the people who could not afford to enter its replacement, the commercial system? Well-to-do, erudite men brawled over the answer to this question, but readers can safely turn to Chapter 5.

5

"Was Once a Science,
Now's a Trade"

Why is it that some people cannot get medication because they cannot afford it? It is easy to assign responsibility to corporations, but there is an even deeper reason why access to medication, including life-saving treatment, is not available to everyone. Take the case of insulin. In 2018, the Kaiser Family Foundation released a study showing that one in four adult Americans with type 1 diabetes had rationed their insulin at least once because they could not afford to buy enough at a time. That figure included people with health insurance. Rationing insulin means taking less than the required dose to stretch your supply longer, and in case you are wondering, yes, that can be fatal. Five years after these data were announced, in 2023, Americans witnessed a flurry of activity at the state, federal, and corporate levels. The Attorney General of California brought suit against the insulin manufacturers "for driving up the cost of the lifesaving drug through unlawful, unfair, and deceptive business practices". A federal law went into effect capping insulin prices for some low-income people using Medicaid, a federal health insurance programme. Two senators proposed a bipartisan bill to cap the price of insulin for everyone, regardless of their insurance status. The three corporations responsible for 90 per cent of manufactured insulin reduced their prices. That is quite a lot of change for one fiscal quarter. These actions did not make insulin available to everyone; they made insulin unavailable to fewer people. But suppose the outrage did not manifest in lawsuits and legislation that focused on insulin as a commodity? In other words, suppose the perceived problem was not the price of insulin but that insulin had a price?

That was the problem facing those practitioners who were commodifying medicinal treatments during the late seventeenth and early eighteenth centuries. They recognized the significance of putting a price on medication: some people would not be able to pay, which could only have a bad outcome. They talked about what should be done, and sometimes argued or came to blows about it. In the end, they knowingly chose to value profit over human health and life and deliberately worked not only to validate and normalize that value but also to render it invisible, a given.

Medical professionals had been balancing the conflicting virtues of getting paid with giving charity to the poor at least since the Middle Ages. After 1650, the dilemma became much more acute as physicians and apothecaries worked to commodify medication. Following outbreaks of smallpox and plague, especially the Great Plague of 1665–6, the issue became urgent. The existing methods for connecting the impoverished with medical treatment were obviously utterly incapable of handling demand during crises, with horrific results: mass graves, bodies in the streets, unchecked contagion. Either professionals would incorporate into the system a method for enabling the poor to get medication, or professionals would continue to build a system that placed profit over access. As Dr Samuel Garth put it in his poem "The Dispensary" (1699), medicine would be either a "Science", dedicated to the acquisition and deployment of knowledge, or a "Trade", dedicated to producing goods and acquiring income. The professionals chose the latter, legitimizing and normalizing the idea of people dying so other people could maximize their income.

A rehearsal for this crisis had taken place in the sixteenth century. When Henry VIII declared himself the head of the Church of England so that he could annul his marriage to Katherine of Aragon and marry Anne Boleyn, he dissolved the monasteries and convents. Cloistered communities tended their own well-being, of course, but for centuries as part of their charitable mission they also provided care to travellers and the local community, especially the impoverished. The Benedictine Rule emphasized care for the sick. Monasteries and convents of all orders had their own physic gardens, which usually would be tended by a monk or nun appointed to be the herbalist. There were cloistered communities all over England in the sixteenth century, so by eliminating them, Henry VIII also eliminated one of the two medical resources available to the vast majority of his subjects. Individual practitioners declined to fill the void that the king had created. It turned out that given the choice between caring for sufferers who could pay and those who could not, professionals preferred the paying kind.

Even if no one had anticipated this thoroughly unsurprising outcome, someone should have warned His Majesty that a tiny number of individuals could not replace an entire network of healthcare. At any rate, Henry VIII was irked. The result was legislation known as the Herbalist's Charter of 1542, officially labelled 34 & 38 Henry VIII c.8. Because "the surgeons admitted [they] will do no cure to any person, but where they shall know to be rewarded with a greater sum or reward than the cure extendeth unto", Parliament authorized local unlicensed, untrained experts to practise medicine "for the ease, comfort, succour, help, relief, and health of the king's poor subjects, inhabitants of this realm, now pained or diseased, or that hereafter shall be pained or diseased". Consequently, the act announced:

> It shall be lawful to every person being the king's subject, having knowledge and experience of the nature of herbs, roots, and waters, or of the operation of the same, by speculation or practice [...] to practise, use, and minister in and to any outward sore, uncome [sic], wound, apostemations, outward swelling, or disease, any herb or herbs, ointments, baths, pultess [sic], and emplaisters, according to their cunning, experience, and knowledge in any of the diseases, sores, and maladies beforesaid, and all other like to the same, or drinks for the stone, strangury, or agues, without suit, vexation, trouble, penalty, or loss of their goods...

The Charter empowered and protected a much larger group than the professionals comprised, which was good, at least in theory. It did not create a system to replace the one that the king had obliterated, which was not so good. In the next century, and using the Scientific Revolution rather than royal prerogative, the professionals eliminated the remaining resource for ordinary people, and like Henry VIII, put nothing in its place.

In effect, the Herbalist's Charter conferred royal approval and legal protection on the practitioners of domestic medicine. In earlier chapters, I showed how domestic medicine was built on and nurtured a set of values, among them that knowledge should be shared and shared without strings or cost, and that people who needed treatment should receive it, also without strings or cost. Women from the top of the social scale down to the poorest woman who could fulfil this role made sure that everyone in the household and their dependents (servants and tenants included) got care. When Anne Clifford, Countess of Dorset, Pembroke and Montgomery heard that her mother was ill in London, she immediately sent a servant with "certain cordials and conserves" to help her mother recover. John Shirley's *The Accomplished Ladies Rich Closet of Rarities: Or, The Ingenious Gentlewoman and Servant Maids Delightfull Companion* (1687) addresses women in every class, but the first chapter, on making medication, is aimed at gentlewomen in particular. It instructs this audience "how to Distill and draw off such Waters from Herbs, and other Cordial matters, as may contribute to the preservation of health, and wherewith a Gentlewoman, being furnished, may be instrumental in saving the Lives, or at least in doing good to her poor Neighbours".

Domestic medicine created and preserved communities. It supported the now-obsolete definition of "family" as the people who lived together in a household, from the lowest stable hand to the master of the house. Under the system of domestic medicine, illness was a communal experience. For healers, domestic knowledge of medications was common knowledge in the sense that women shared it and expected to share it. As I discussed previously, recipe books connected women across time and space: across backyard fences, villages, counties, regions, England, Scotland, classes, and generations. No one owned that recipe for an effective eye wash any more than she owned that recipe for pear tart. It was so much a part of culture that literature could use the interaction as a marker of virtue. In the narrative *The Wandering Beauty* (1698), the heroine Arabella sees a baby with "bad" eyes and offers a cure to the mother (it works).

Nor did the ill or injured suffer alone. They were attended by members of the household and, when necessary or helpful, by members of the extended family or community. Pregnancy and childbirth were viewed the same way, as one long experience involving women of the family and community. The welfare of a family was affected by the health and well-being of its members – the death of a father or mother could send the household plummeting into destitution – but it was also a question of ethos. One person's illness was everyone's concern. Domestic medicine's values were cultural; they were organizing principles of social practice and structure for everyone. It was the way people understood what it meant to be in or out of health. If a labourer complained to his buddy that he had a strange rash on his arms, the buddy would ask if he had shown the rash to his wife. That is just what one did.

It was not a form of charity. True, a great deal was made of charity in Christian Europe. Individuals were exhorted to be charitable. However, charity involves a vertical distribution of resources, one in which virtue accrues to the resource-rich individual choosing to part with some of those resources. Domestic medicine involved the lateral circulation of resources. Sick people needed treatment; sick people received it. Recipes passed from equal to equal to empower the recipient. If you do not think of knowledge as private property, or of something like burn salve as a commodity, you do not worry about losing or gaining property from giving or withholding it. When it came to medicinal recipes, sharing was more than a norm. It was an expectation and an obligation.

In fact, "I think myself obliged to send it" is how Mary

Huntington explained sending a medicinal recipe to John Locke on 5 January 1699. That is the John Locke who wrote *An Essay Concerning Human Understanding* and *Two Treatises on Government*; coined the phrase *tabula rasa*; helped draft the English Bill of Rights; provided the foundation for the American Declaration of Independence, Bill of Rights, and Constitution; and who was himself a practising physician with his own recipes. Mary Huntington was the wife of one of Locke's old friends, and her letter was precipitated by the news that he was having breathing problems. She had the recipe for a good cure and "as I desire your Health (which I most heartily doe)," she wrote, "I think myself obligd to send it". Properly deferential to "the Great Philosopher and Physician of the Age", as she called him, Huntington began her letter by noting that Locke did not need her advice, writing, "We have a proverb, sir, of carrying coals to New Castle," but she continued defiantly, "and I am now going to act it." And although she ended her letter with an apology – "If I be impertinent, I beg your pardon" – the apology itself ended with "but if it should have the blessed effect I wish it, then I'll not ask you to forgive". Considering that she was confident in the cure, and considering that she sent the letter, she could not have been too worried.* In fact, her recipe might well have eased the philosopher's respiratory distress. It called for brewing liquorice and figs in "Spring Water" and drinking the mixture "in a wine glass". Laboratory analyses have confirmed that liquorice, an

* Huntington did not lack for determination. When her husband Robert Huntington died in 1701, she badgered his colleague, the distinguished Oxford scholar Thomas Smith, to write his biography until Smith caved in to her demand in 1704.

ancient treatment for chest congestion and phlegm, really does work. Figs appear to have anti-inflammatory effects.

Domestic medicine was not a system for treating the penurious ill. It could not provide care for everyone, but it could offer a set of values. Instead, the hierarchical thinking of charity shaped the initial organized responses to the sick, impoverished population. The first earnest, effective governmental attempt to create a system for tending to the medical needs of the impecunious came in the form of the Poor Law of 1601, often referred to now as the Old Poor Law. It was aimed at people who were truly desperate, who had nothing – no home, no food, no family or acquaintance to help, no income, nothing. Until the nineteenth century, the Poor Law required all parishes, urban and rural, to collect fees from residents to pay for food, employment, and medical care for those unable to provide it for themselves. Parishes paid weekly pensions to residents who proved need.

Urban parish overseers sometimes established a workhouse, where desperate indigent people could live and work, perhaps get some education, and obtain medical care supplied by a hired surgeon and apothecary. The first workhouse in Birmingham, established in 1727, employed three practitioners. Workhouses were generally inundated with people desperate for treatment. It was a significant financial burden for a parish, and the cost skyrocketed over the seventeenth and eighteenth centuries. Additionally, each parish negotiated (or did not) the price of medical care and specifically of medications. When the RCP offered to supply medications for the price of what it cost to make them for one year to the Bishopsgate Street workhouse, the Worshipful Society of Apothecaries outbid them by proposing to supply medications to the workhouse for three years for free. It was a great deal for the apothecaries, who burnished their reputation, and for the Bishopsgate parish officials – but only for the Bishopsgate parish officials.

Beyond the Poor Law, there was a patchwork of methods for providing care for those who could not afford it. After the dissolution of the monasteries, church hospitals sometimes continued under different management. In Gloucester, for instance, three hospitals – St Bartholomew, St Margaret, and St Mary Magdalen – came under the control of the city government, called the Corporation. The same happened to St Catherine's in York. Small parishes and the governments of smaller cities and towns often hired a medical man to treat the indigent in addition to his regular business, or arranged to recompense certain providers (apothecary, surgeon, or physician) for doing so. In the parish of Alfriston, one Richard Alcorn was reimbursed by "Mr Brooks" for, among other things, "17 powders", "8 papers", "18 Drops", "six bolusses", and "an oyly mixture", as well as for "opening the swelling on Martha Asten's hand" and a "visit & bleeding" for Thomas King. Especially in the hinterlands, "wise women" were paid to handle lesser ailments and injuries. Occasionally, individuals in towns and cities across England formed "friendly societies", which functioned like health insurance: everyone in the group paid in, and when someone needed care, the fees would be paid from the kitty. Further afield in the countryside, it was not uncommon for landowners and clergymen to obtain medical training so they could care for the neighbourhood.

And of course, there was private charity. To borrow from Tennessee Williams, the impecunious ill had always depended on the kindness of strangers. In towns outside London, professionals sometimes attended individuals without fee, while in some places, apothecaries banded together to provide care or medications for the local poor. Guilds might take charitable action. St Thomas's and Trinity Hospitals in York wound up in guild hands. In London, St Bartholomew's Hospital, established in 1123 and famously known as St Bart's, and St Thomas' Hospital, established in 1173, were still operating on their own in the seventeenth and eighteenth centuries. Britain also was dotted with establishments founded by very wealthy individuals: hospitals to care for those unable to care for themselves – usually because of age, poverty, or both – and almshouses for the homeless. These institutions tended to be small, averaging about twelve residents. Thomas Oken's Charity in Warwick maintained almshouses for six people. The Wolborough Feoffees and Widows' Charity in Devon provided housing and a tiny annual stipend to a small number of widows; the Yerbury Almshouse in Wiltshire housed six widows and provided each just under four pounds per year. Thomas Gouge devoted an entire book, *The Surest & Safest Way of Thriving* (1673), to biographical sketches of self-made men who exemplified charity, such as Mr John Walter, a draper, who built two almshouses in London. As Lady Lucy says to Sir James Courtly in Susanna Centlivre's play *The Basset Table* (1705), "That Superfluity of good Manners, Sir James, would do better Converted into Charity; this town abounds with objects."

Private or "voluntary" hospitals, funded by a group of benefactors paying an annual subscription, also began to appear in the early eighteenth century. The anonymous author of *The Present State of Physick and Surgery in London* (1701) noted the endless demand for this kind of succour: "The Governours of our Hospitals, who give their charity in directing the charity of the Foundings in their respective Houses, where the health of some Hundreds is provided for, cannot observe the calamities of many Thousands without concern, and their Endeavour to promote their relief." In London, for example, the Westminster

Infirmary was founded in 1719, Guy's Hospital in 1724, St George's Hospital in 1733, and the London Hospital in 1740. The Glasgow Town Infirmary opened in 1733, staffed by volunteers from the local Faculty of Physicians and Surgeons; benefactors and staff physicians paid apothecaries for the necessary medications. Bristol opened its voluntary hospital in 1737 with a full medical staff, and Aberdeen opened its first voluntary hospital in 1740.

Organizations like these had a limited impact, even the Charterhouse, the most significant private charity in the British Isles. It was established in London in 1611 by Thomas Sutton, who was outrageously wealthy, spectacularly connected, and extremely well liked. Sutton bequeathed a London property, Howard House, and an ample endowment to establish and maintain a combined retirement home (hospital) and boys' school. Sutton directed that eighty pensioners and forty students should be provided for, an outrageous number in comparison with most charitable hospitals, almshouses, and schools of the day, and he left a magnificent real estate portfolio to fund it. The education provided for the boys was famously good. In the short play *The Triumvirate* (1719), the character Harlequin assures his friend Scaramouche that "I have Latin and Greek enough left, since I was a scholar at the Charterhouse." To qualify for the retirement side, pensioners had to be old and fallen into poverty through no fault of their own after they "had lived active and useful lives in conditions of prosperity and comfort". To be admitted, a pensioner had to be nominated and accepted for admission by a Board of Governors. He was guaranteed a fully furnished private room, board, and medical care and could remain at the Charterhouse until death.

Sutton's network and awe-inspiring wealth ensured that his institution was well provisioned. In addition to the reigning monarch, his or her spouse, and the heir to the throne, the Charterhouse's Board of Governors included the Archbishop of Canterbury, the Bishop of London, the Bishop of Ely, three other bishops, the Lord Chancellor, the Lord Chief Justice, and four privy councillors. The large staff included a physician, who received a salary and a house on the grounds, and a gardener, whose responsibilities included providing vegetables and fruit for the kitchen, and herbal plants for making medicaments. The physic garden must have markedly alleviated expenses because when the soil became irremediably overworked in 1731, bills for medication (not to mention vegetables and fruit) increased.

Medication was an imposing expense for every charitable organization; considering that the Charterhouse's two populations were schoolboys and older, infirm men, it was probably formidable. Like the governors of other institutions, such as the Westminster Infirmary, the governors of the Charterhouse determinedly attempted to control the cost of medication by harassing the apothecaries who required payment. Between 1700 and 1712, the board challenged every set of bills submitted to them by James Petiver, apothecary to the Charterhouse. Petiver was an impressive figure in his own right, not only a successful apothecary but also a Fellow of the Royal Society and one of the most prominent botanists of his day. The RCP had written in strong support of his appointment. Nevertheless, the governors did not trust him. Petiver recorded that on 12 July 1710, William Hempson, the registrar of the Charterhouse, reported that in response to Petiver's petition to be reimbursed for the medication he made, the governors "referr'd it to their next Meeting in full Assembly & Order'd y^t in ye meantime Dr Goodall shall examine ye bills & give his opinion as to ye Reasonableness of ye Price of ye Physick".

To his credit, the physician of the Charterhouse always supported Petiver. In 1703, Dr Charles Goodall held the post, and on 29 February of that year he wrote the governors that "I have compared these Bills with those directed by me & am satisfied they were accordingly delivered." His letters all said the same thing and almost in the same words. March 1708: "I have Carefully lookt [*sic*] over yᵉ Prices of Mʳ Petiver Medicines relating to yᵉ Pentioners of yᵉ Charterhouse & do find them valued at lower rates then to his other Patients & lower then the late Apothecary Mʳ Holme did rate some of them," while in 1709, Goodall wrote only "The Truth of these Cases I do attest & know." More was required of him in 1712, after the governors balked. "May it please yʳ Honours," he wrote on 14 June, "There having for these 2 or 3 Years last past been several very Malignant & Dangerous Fevers & Small Pox in this House & a great Number of yᵉ Schollars infected with yᵉ Itch," as well as "Violent Convulsive Collics, Stranguries" and "other stubborn Cases too long to trouble your Honours with", it was logical that James Petiver's bills should be higher than expected. In 1712, Petiver recorded in his notes for the Charterhouse that Goodall had told the governors that Petiver's bills were fair, and that Goodall had reiterated that he, Petiver, was charging the same or even slightly less than the previous apothecary to the Charterhouse, Mr Holme.

The next physician to the Charterhouse, Dr Henry Levett, also attested to Petiver's honesty. After reviewing the records, Levett confirmed that Petiver had done exactly what Dr Goodall had ordered and at a reasonable price. As for his own prescriptions, Levett was "satisfied yt all these medecines were delivered with Care & according to Order", a phrase he used again in his June 1714 letter to the governors. The governors were fortunate, however ungrateful they may have appeared.

The Westminster Infirmary went through long periods of difficulty with apothecaries; during one tumultuous eighteen-month period, it hired and fired three of them.

The Charterhouse's governors were also probably aware that the medical staff of charitable institutions sometimes used the institution that employed them for their own ends. After all, private and voluntary hospitals usually were staffed with professional men who knew each other. Alexander Stuart and William Wasey, future distinguished Fellows of the RCP, shared a casebook for their patients at the Westminster Infirmary between 1723 and 1724. Samuel Garth sent several insolvent, ailing patients to Hans Sloane bearing notes requesting Sloane's help in procuring treatment for them. In one, Garth asked Sloane to "gett this poor Maid" into Christ Church Hospital; in another, he wrote, "I beg you'll be so charitable to gett this poor woman into some Hospital." (Garth was not always so kindly. He asked another friend to "doe [sic] an action of Charity" for the bearer of the letter, "this miserable slut".) A pair of letters in the Sloane Manuscripts at the British Library even shows one physician to the Charterhouse daringly attempting to evade the dragonish governors.

The letters record an exchange between one Mary Burley of Reading and the physician Dr Henry Levett about her son's condition.* The reply letter was penned by "J.P.", that is, James Petiver, from his shop in Aldersgate Street, under the sign of the White Cross. The correspondence centred around Mrs Burley's son, who had suffered a terrible blow to the head some

* As of this writing, the British Library identifies the physician as Dr Charles Goodall, but this is an error. The letters are dated July 1713, when Goodall had been dead almost a year. Either he was a miraculously dedicated physician or the addressee is Dr Henry Levett, his successor.

weeks earlier. On the morning of the letter, his anxious mother wrote: "My Son have this day A very bad fit; being taken in yᵉ morning with a great payn whare [*sic*] yᵉ blow was, & so sick & faint with it that he could neither stand nor spake for a very Considerable time." She reported that she "gave him to open him as soon as he could take it an infusion of rubarb [*sic*] & lickirish; and when that came to work with him; he had ease". That is, Mrs Burley gave her son a drink of rhubarb and liquorice, and once he began to feel its effects, his condition improved, although "he is very weake and faint, as he allwaies is after such a fit". The effects were in the purgative line: pow-dered rhubarb was used as a laxative (and yes, it works), while liquorice (as Mary Huntington knew) is an expectorant. Mrs Burley may have administered it as a liquid or "juice" given her term "infusion". Apparently young Burley's caregivers thought to purge him until his brain injury healed. Unsurprisingly, it was not working.

Mrs Burley's appeal did not come out of the blue. Levett had treated her son for the injury before. As she explained, "I was in hopes to have hard [*sic*] from you before now I rit by last Tuesdays post to acquaint you how he was" because "he is faln back for want of proper things to take", that is, Levett's prescribed medicine. As politely as urgency allowed, she entreated him for a reply: "I bag [*sic*] I may hear from you as soon as posable." To ensure that she received any medication sent from London, Mrs Burley included detailed instructions for posting it: "yᵉ Apothecary may direct to me in London strete in Reading; yᵉ Reading coach goes out every day from yᵉ b[*illeg*] & Tun inn in flete strete & from yᵉ white hors [*sic*] in fleet strete." The illegible word and repetition hint at her distress.

James Petiver wrote the answer:

Madam

I sent yr Letter to ye Dr as you directed, who has prescribed him a purging Electuary to take 2 or 3 times a weeke if Costive, & yt he would have him come up to Town if worse, because he cannot prescribe so well for him at yt distance Nor will ye Governors pay for ye Physick or Advice wch is given out of House In ye interim wt he has now & later Ordered is very proper for him to take & I heartily wish him ye good Effect of them who am

    Madam yr humble servt to command

J. P.

Aldersgate Street [*illeg*] Juli [*sic*] 7. 1713

It was not uncommon for a physician to entrust the apothecary with a prescription and the explanation for how the patient should take it. Petiver's extra reassurance ("In ye interim wt he has now & later Ordered is very proper for him to take") demonstrates familiarity with physicians' recipes and a respectful relationship with the doctor, not to mention consideration for a terrified mother. Mrs Burley also might have found relief in having used domestic medicine to do what the professional medicine would do ("costive" was a medical term for constipated). That she relied on a London physician's advice and medication, however, shows how much professional medicine had gained on domestic medicine by 1713.

There is one odd element to this letter. The recommendation to bring Mrs Burley's son to London to be treated in person, logical as it sounds to someone reading this book, was not a given. (Prescribing without examining the patient will come up again in Chapter 7 because quacks encouraged patients to self-diagnose.) But Levett had two motives for asking Mrs Burley to come to "Town": to see the patient for himself and to get the Charterhouse to pay for the treatment. Hence the warning about the "Governors", the explanation of paying for a consultation and the prescription, and the mention of the "House".

These methods of connecting impoverished sick people with medication were insufficient to the demand. The more that access to medication depended on money, the less access there was. Looked at one way, this was and is a significant moral issue. A real number of people who could not pay for treatment became sick, maimed, debilitated, or dead. A system that withheld medication from those who needed but could not afford it was a system that accepted the deaths of some people so other people could profit. This was the moment, as they were commodifying medication and creating an economy around it, for professionals to devise a mechanism ensuring access for everyone. They knew it, too. They even discussed what to do, as well as whether to do anything, to ensure everyone could get treatment. In the end, they let the opportunity go by. Instead, and rather reminiscently of Henry VIII, medication's commodifiers let the hierarchical practices and values of charity continue as inadequate suppliers of care for those who could not afford it. It was perfectly fine if those who already had acquired more and more, and if those who had not did not experience a change in situation.

Just who were "the poor"? It is a capacious category. "Poor" describes people who had nothing: no means of feeding, sheltering, or clothing themselves. Generally lumped in with this group was another, who might be called the "solvent poor". People in this class worked jobs, resided somewhere, and had food and clothing to some extent. When the Scientific

Revolution arrived in the British Isles, the solvent poor also had domestic medicine and, in some areas, apothecaries. Apothecaries had a pretty good reputation for caring about the lower classes and the indigent. For one thing, they did not leave during outbreaks of plague or smallpox the way the wealthy (and physicians) did. During the Great Plague of 1665–6, apothecaries remained in London, while all but a very few physicians left. True, an apothecary's living was inseparable from his shop and its stock, neither of which could be packed up and moved, so it is perhaps not so surprising that they stayed put. But the shop also bound its apothecary to a neighbourhood, making him a member of a community with established relationships with his customers. These connections were particularly strong in rural areas or small towns. In Coventry, for example, the apothecary Abel Brooksby was elected mayor in 1672. Apothecaries had some ability as well as incentive to match the price of their medications to their customers, to fit the micro-market that they served. It was not unusual for apothecaries to provide medication on credit, in fact. That economic flexibility ensured that more people could get medication and nurtured a market for the apothecary's goods. A poor family might not be able to afford a physician, but they might be able to afford an apothecary.

Physicians did not have a reputation for caring about the impecunious. Like the members of other guilds, physicians everywhere depended on income generated as individuals; in their case, consulting and diagnosing fees. The more famous the physician, the higher his fees. In the late seventeenth century, the pre-eminent London physician Dr John Radcliffe charged up to £20 for a house call, roughly equivalent to £4,000–£6,000 in the early twenty-first century. At the time, 94 per cent of the population earned less than £100 per year; less than 1 per cent had more than £500, the minimum required to maintain a genteel lifestyle. When the protagonist of *Physick Lies A-Bleeding* (1697), Mr Trueman, is handed the bill for his wife's treatment, he grumbles that "I do suppose to be genteel, I must give you a crown." Plague or smallpox could erupt in any urban centre (the plague could even be tracked as it swept across continents, nations, or cities), and when epidemics struck, those who could afford to, fled. Of course, those who fled tended to be the people who could afford a physician, so unsurprisingly, physicians fled with them. Individual physicians might have a reputation for generosity, compassion, or benevolence, but between the paucity of physicians overall, the absence of physicians at times of crisis, and their being expensive, the stereotypical doctor was not imbued with generosity or fellow feeling.

So what happened? Scotland offers examples of what might have been. One of the first actions taken by the Royal College of Physicians of Edinburgh, founded in 1681, was to establish a dispensary that provided free medical care to the poor. Initially that treatment took the form of house calls, but after 1705 it took the form of an outpatient clinic at the Edinburgh College's headquarters. From its inception, the Edinburgh Dispensary was staffed by two physicians elected by their peers to serve for twelve months. It was funded at first by fines paid by the membership for assorted infractions, and then as part of the College's entrance fee. Medication still had to be paid for, but not by those receiving it. The Glasgow Town Infirmary technically belonged with its attached workhouse, but in fact it functioned separately. Volunteers from the city's professional medical organization staffed both infirmary and workhouse, and an apothecary provided the medications out of his own pocket. Sometime around 1740 this position was given a salary, some of which was to cover the price of materia medica and whatever else was needed.

These were not the only ways that professionals responded to the question of whether the poor could get access to medicinal drugs if they could not pay for them, nor were any of them the one that became embedded in the medication economy. For that, we can turn to London: the story of events there is the tale in miniature. In London, different guilds responded quite differently to the Scientific Revolution. The Worshipful Society of Apothecaries as an entity adopted the Scientific Revolution's principles, methods, and fascination with tools and technology. The guild saw an opportunity to reduce the cost of materia medica for its members and to create reliable sources of revenue. From its inception in 1611, the Worshipful Society of Apothecaries recognized that controlling a member's expenses increased the profit margin, and also made it economically more feasible for him to attend the lower classes. The London guild built a laboratory at its guildhall, possibly as early as 1623, to train apprentices and to make in bulk supplies for members to buy from the guild and sell in their shops.

After the Great Fire burned it down and two-thirds of London with it, the Worshipful Society of Apothecaries rebuilt the hall with a large laboratory in the cellar and commercial ambitions in mind. Opened for business in 1672, the laboratory quickly became, as historian Anna Simmons puts it, "one of the earliest sites in England for large-scale drug manufacture". In this way, the laboratory suppressed the cost of medicinal simples and compounds, and also enriched the guild as a body. By 1680–1, the laboratory was so useful that it was turning a 30 per cent profit. In large part because of their manufacturing capacity and the consistency of their mass-produced compounds, in 1703 the Worshipful Society of Apothecaries won a highly lucrative contract to supply the medications for the Royal Navy, a contract that the organization held for more than a century. In 1673, the London apothecaries acquired what became known as the Chelsea Physic Garden and started growing many of the necessary herbaceous, arboreal, and floral ingredients for their medications, giving them even more control over prices. This strategy also benefited apothecaries outside London. They too could buy the complex and chemical compounds from the London guild's laboratory, expanding their materials, following the latest medical developments, and keeping their prices suited to their own local markets.

While the apothecaries were making use of the Scientific Revolution's emphasis on technology and innovation to establish an economic foundation for their guild, the physicians were not. When they rebuilt their headquarters at Warwick Lane after the Great Fire of London in 1666, they added an anatomy theatre that became internationally famous but was not designed to produce revenue for the College. Generally speaking, proponents of the Scientific Revolution gathered at the Royal Society once it was chartered by Charles II in 1662. Some of the Royal Society's Fellows were apothecaries: James Petiver, for one, and Isaac Rand, who directed the Chelsea Physic Garden. A significant number of the Fellows belonged to the RCP, such as Hans Sloane, Alexander Stuart, and Richard Mead. But on the whole, the RCP was slower to accept the Scientific Revolution and the medical revolution that came with it. Warwick Lane was not the place to hear a paper on the most recent anatomical discovery.

Broadly speaking, then, the primary provider of medication (apothecaries) and the two main providers of treatment (apothecaries and physicians) did not engage in the kind of creative thinking that might counterbalance their efforts to take full control of medication and to eradicate domestic medicine. The literature of the time shows that when push came to shove by the end of the seventeenth century, neither group was willing to give up any way of making money, even in the interests of saving lives or ameliorating suffering. Neither group was willing to risk any future part of the medication market, either; they directed the conversation and shaped the vocabulary around the question of the poor's access to medication, hiding the real values under their choices and deflecting attention from those values and choices onto other actors.

Until the 1690s, the apothecaries of London and individual benevolent physicians were left to treat ailing people who could not pay a doctor's fees. Before that, the RCP had made several feints at an outpatient clinic, which they called the Dispensary, to be staffed twice weekly with physicians donating their time. Discussions in the College in 1670 and 1687 petered out, the latter producing only an order to the Fellows to donate some of their time to treating the poor, an edict that changed no one's behaviour. When the College brought up the idea in 1675, they approached the apothecaries about providing the medications for the Dispensary, but their tone was not conducive to partnership: "Not doubting but that the Company of the Apothecaries will suitably comply with our just and real intention and désigné of serving the public in affording medicines pre scribed by us to such poor at rates answerable to the lowness of their condition." Unsurprisingly, the apothecaries did not choose to suitably comply, and so yet again nothing happened. Inertia is remarkably comfortable when the money is coming in.

Things changed at the RCP in the mid-1690s. In 1695, the governing committee of the RCP convened a subcommittee, called the Committee of Medicines, to investigate the feasibility of the Dispensary. On 30 October 1696, the minutes for

(*above left*) "Haw-thorn" (no. 149) from Elizabeth Blackwell's *A Curious Herbal*, vol. I, 1737. Note the seeds and cross-sections, as well as Blackwell's signature (bottom left): "Eliz. Blackwell delin. sculpt. et Pinxit" (Elizabeth Blackwell outlined, etched, and painted this image).

(*above right*) Professionals and their bad reputation. Thomas Rowlandson or follower, *Death as an apothecary's assistant.*

(*left*) Lady Elizabeth Grey, Countess of Kent, painted by Paul van Somer, *c.* 1619. The Countess is likely wearing mourning for Queen Anne, with whom she was close. The low neckline was fashionable for women of the time.

To cure a sore Breast.

R. Empl: de Ranis cum Ðio ℥ß
somnj ferj thebai. ʒij.
Take of yᵉ Implaister of Frogs wᵗʰ Mercury two Ounces.
Opium ʒij mix ym & make a plaister for yᵉ Breast.

To pickle a Hamm.
If it be a large one take a pound & half of brown
suger & rub it all over very well wᵗʰ itt & lett it lye till
ᵉ suger is all melted upon it, yⁿ take half a pound of
salt Petre & pound it very smale & 3 peny worth of
Chochonell pound'd, mix ym well together & add what
quantity of common Salt you think fit & rub yᵉ Ham
very well wᵗʰ it & let it lye till yᵉ salt is all dissolv'd,
n hang it up in yᵉ Chimny for 3 weeks or a month. yᵉ
quantity if yᵉ Hamm's are not large may serve to pickle
two, & little ones are better yⁿ great ones; yᵉ Pickle yᵗ
omes from ym is very good to pickle Tongues in or Beefe,
d will make either very red & tender.

An approv'd medicine to procure deliverance of a dead
Take 3 Dragon Roots & stamp ym & divide & bi
yᵉ hollow of yᵉ feet:
To procure Throws; Twelve Spoonfull of Claret
Wine, five drops of Oyle of Nutmegs & a little Suger
Travell In Travell; Take five Spoonfuls of red Wee
it wᵗʰ London Treacle yᵉ quantity of a Walnutt &
it in Travell, it opens yᵉ Body.
To try yᵉ Body; Let yᵉ Body be tryed wᵗʰ
of sweet Almonds by yᵉ Midwife.

To make Gooseberry Wine.
When Gooseberrys are full ripe break ym like Ap
for Syder & to every bushell of Gooseberrys 2 Gallons
boild water wᵗʰ cold & press it hard & put in two p
of fourpeny Suger into every Gallon of Liquor & strain
very well & it stand in an open vessell two or three
& take off yᵉ Scum & yⁿ Tun it into yᵗ vessell & not
it till done fermenting & yⁿ keep it half a year, & yᵉ
For a Cow troubled wᵗʰ yᵉ Tardy.
One pound of Raisons, two peny worth of Cana, half a
of Sallet Oyle, two peny worth of hony;

Early eighteenth-century recipe book, owner unknown.
The recipes on these pages (for nursing, giving birth, cooking, and livestock)
demonstrate the variety characteristic of these books.

Life-sized family portrait of Sir Kenelm Digby (1603–1665) with his wife Venetia
Stanley (1600–1633) and their two eldest sons, Kenelm (1625–1648) and John
(1627–1673), by Anthony van Dyck, c. 1632. Venetia might have been pregnant when
she sat for this painting, because she gave birth to George in January 1633.

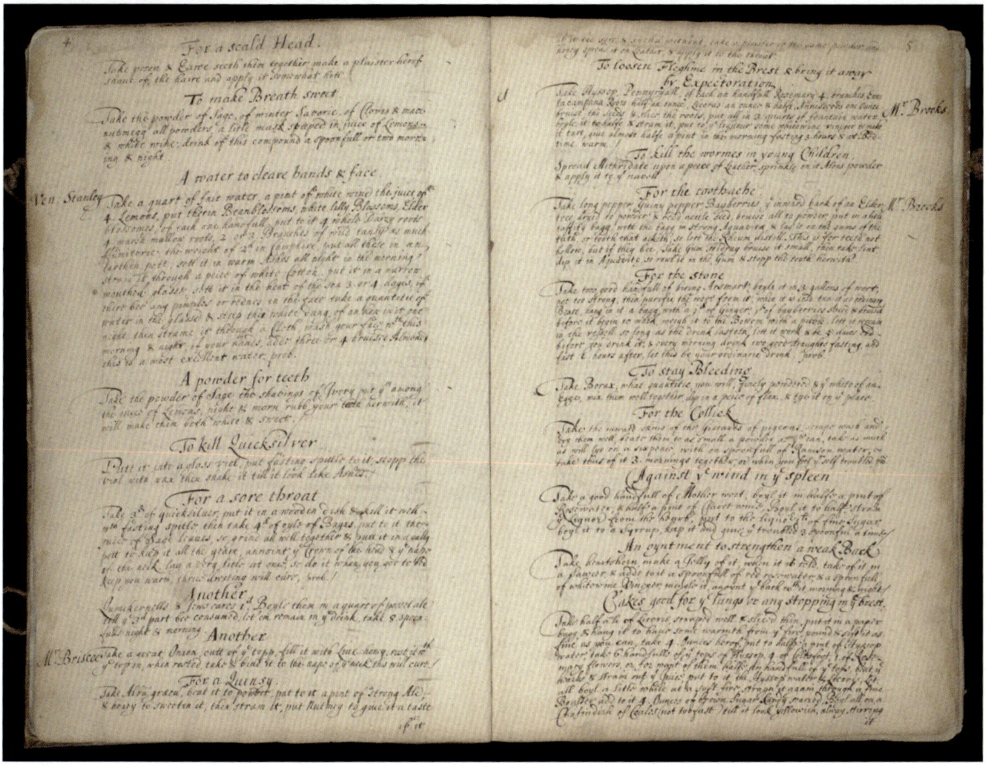

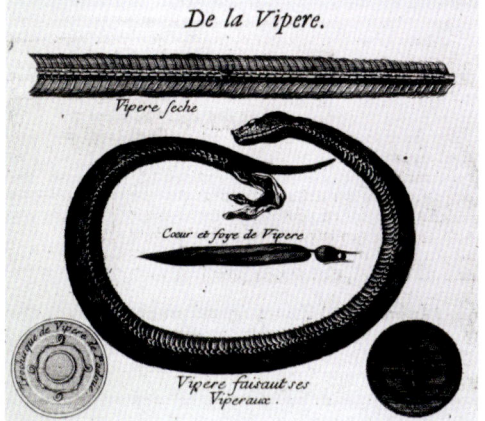

Venetia Stanley's recipe for face wash (left side). Seventeenth-century recipe book, owner unknown.

"De la Vipère" from *Histoire générale des drogues* by Pierre Pomet (1694).

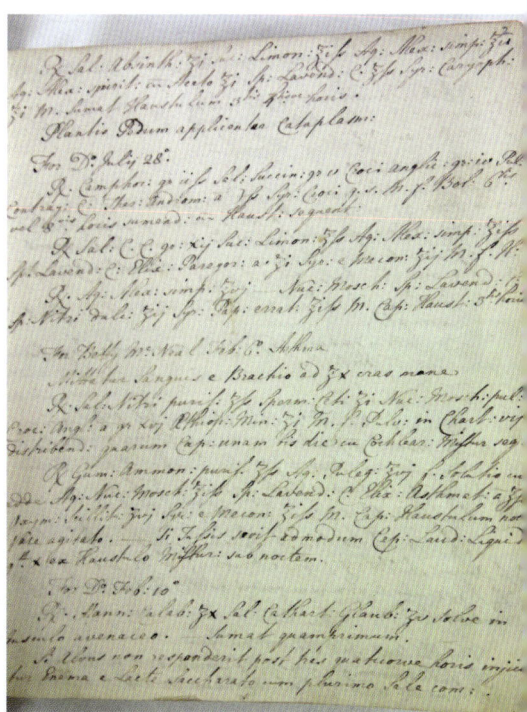

A page from a medical notebook with Dr John Huxham's prescriptions for several patients. When one set of prescriptions fail, the physician tries another. Mid-eighteenth century.

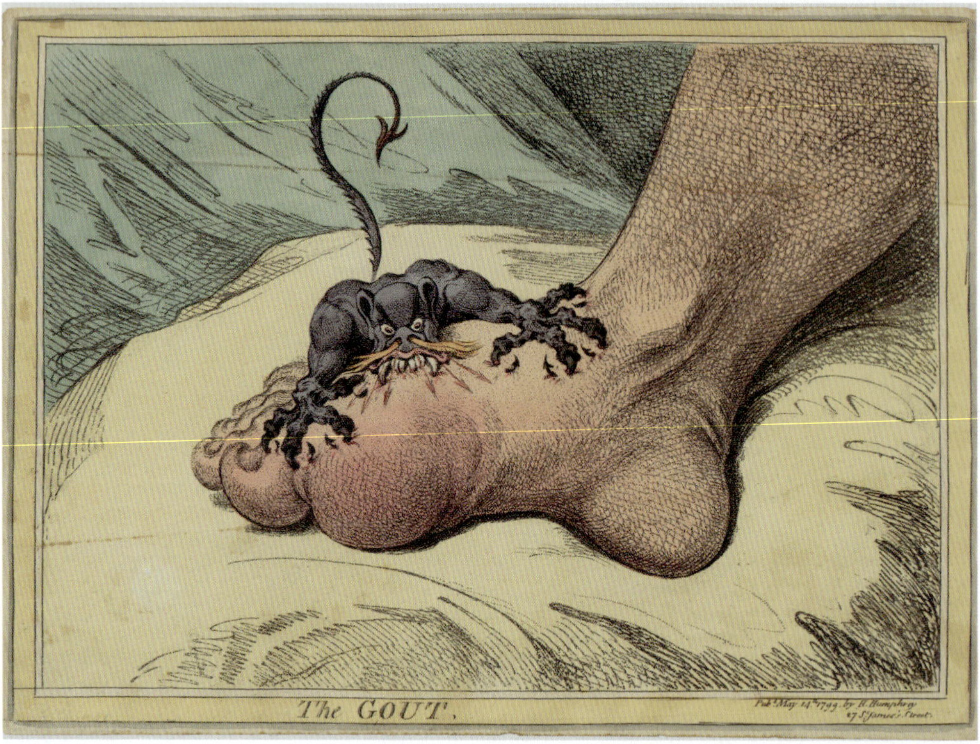

The Gout by James Gillray (1799).

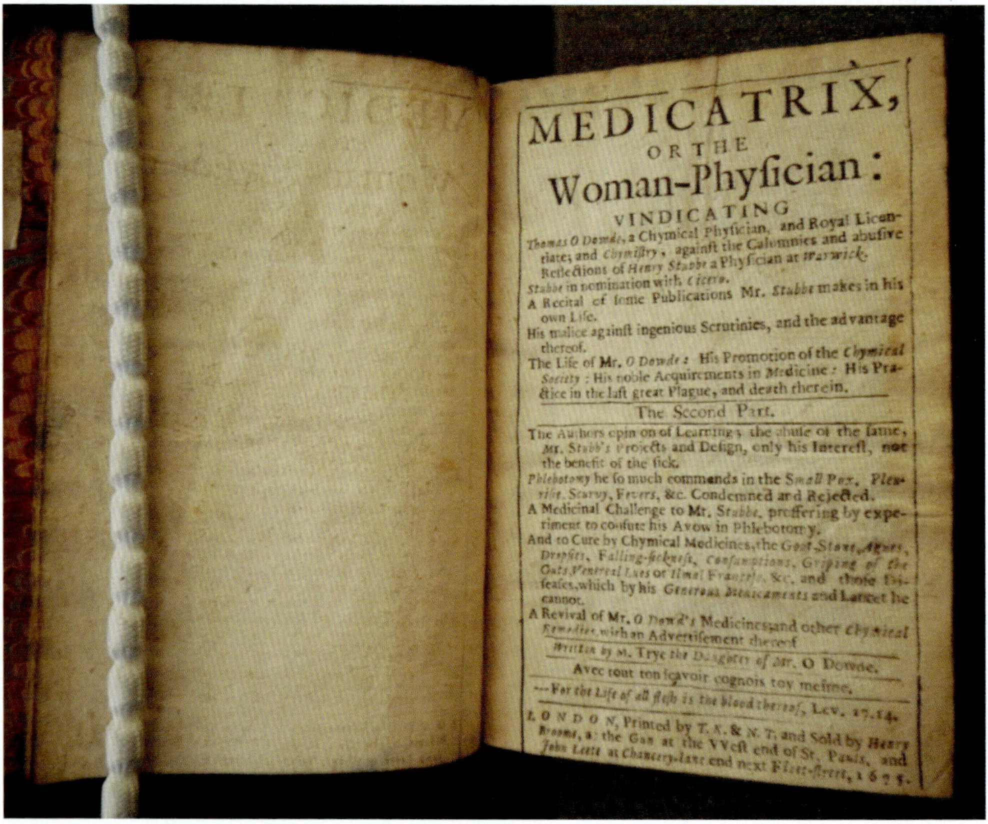

MEDICATRIX,
OR THE
Woman-Physician:
VINDICATING
Thomas O Dowde, a Chymical Physician, and Royal Licen-
tiate; and *Chymistry*, against the Calumnies and abusive
Reflections of *Henry Stubbe* a Physician at *Warwick*.
Stubbe in nomination with *Cicero*.
A Recital of some Publications Mr. *Stubbe* makes in his
own Life.
His malice against ingenious Scrutinies, and the advantage
thereof.
The Life of Mr. *O Dowde*: His Promotion of the *Chymical
Society*: His noble Acquirements in Medicine: His Pra-
ctice in the last great Plague, and death therein.

The Second Part.

The Authors opinion of Learning, the abuse of the same,
Mr. *Stubb's* Projects and Design, only his Interest, not
the benefit of the sick.
Phlebotomy he so much commends in the *Small Pox*, *Pleu-
risie*, *Scurvy*, *Fevers*, &c. Condemned and Rejected.
A Medicinal Challenge to Mr. *Stubbe*, proffering by expe-
riment to confute his Avow in Phlebotomy.
And to Cure by Chymical Medicines, the *Gout*, *Stone*, *Agues*,
Dropsies, Falling-sickness, *Consumptions*, *Griping of the
Guts*, *Venereal Lues* or *Ilmal Francese*, &c. and those Di-
seases, which by his *Generous Medicaments* and Lancet he
cannot.
A Revival of Mr. *O Dowd's* Medicines, and other *Chymical
Remedies*, with an Advertisement thereof.

Written by M. Trye the Daughter of Mr. O Dowde.

Avec tout ton sçavoir cognois toy mesme.

---*For the Life of all flesh is the blood thereof*, Lev. 17.14.

L O N D O N, Printed by T. R. & N. T. and Sold by *Henry
Broome*, at the Gun at the West end of St. Pauls, and
John Leete at *Chancery-lane* end next *Fleet-street*, 1675.

Mary Trye's *Medicatrix* (1675) prepared for a twenty-first-century reader.

A woman's numerous domestic responsibilities. John Shirley, Frontispiece to *The accomplished ladies rich closet of rarities* (1687).

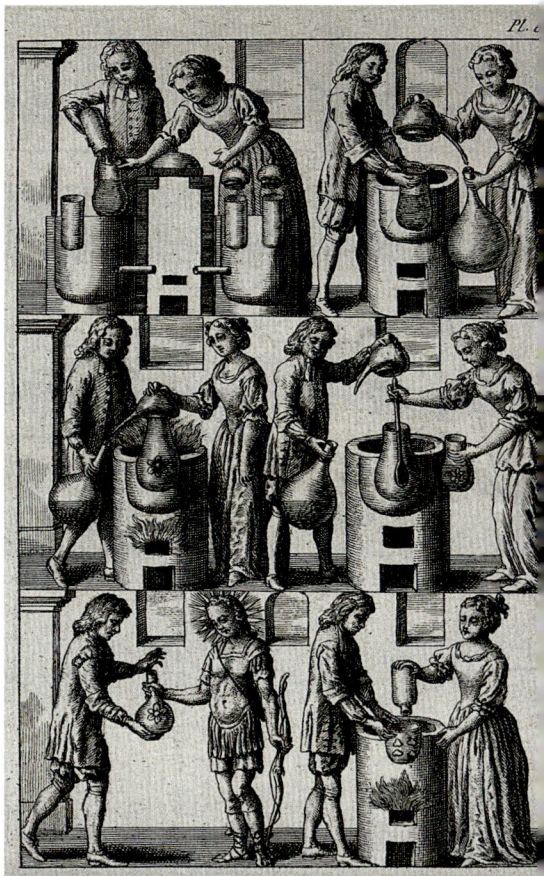

Seventeenth-century illustration showing the well-known overlap in women and men's knowledge and skills. *A man and a woman demonstrating the process of fermentation and distillation in alchemy.*

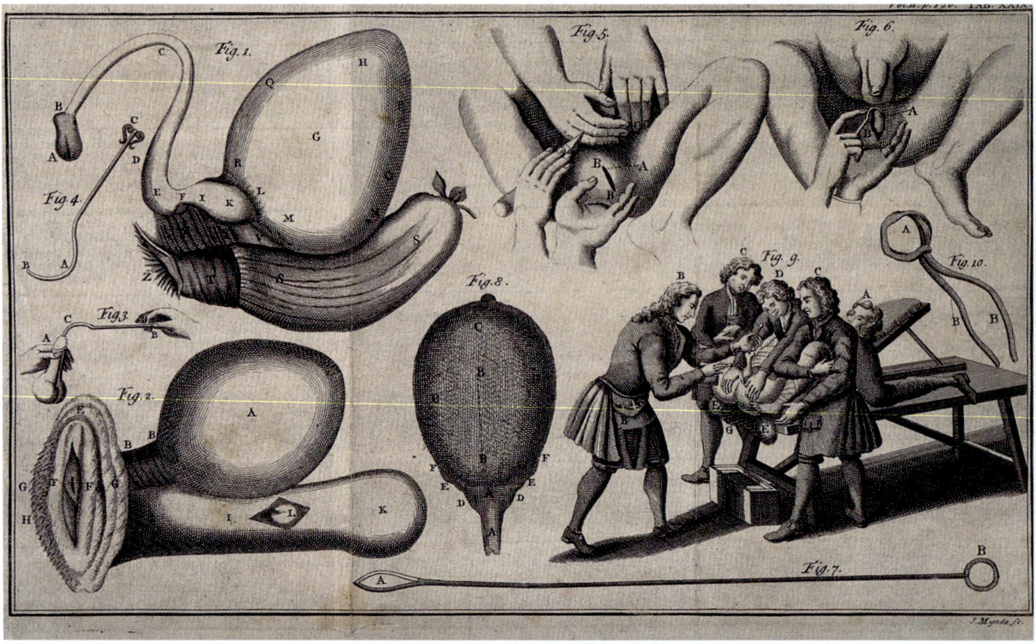

(*above*) A surgeon performing a lithotomy on a patient who is being restrained by three assistants, with five other anatomical illustrations by J. Mynde (1743).

The Company of Undertakers by William Hogarth (1736).

After Pietro Mainoto, *The apothecary making up a prescription, his wife holds a recipe for him…* (Eighteenth century).

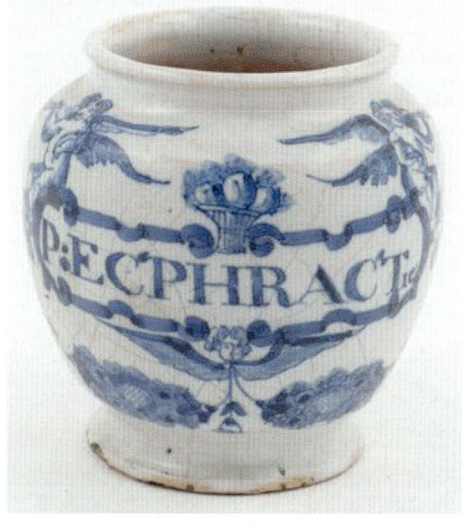

English delftware drug jar with the coat of arms of the Worshipful Society of Apothecaries. First known image of Apollo, God of Healing, on an apothecary jar. In English, the Society's motto reads: "I am spoken of all over the world as the one who brings help."

Small jar for an apothecary shop to store ecphractic (laxative) pills (1700–1740).

Interior of a pharmaceutical laboratory (1747). The chemist and assistant
work in the well-equipped back room. The shop where their products
are sold is visible through the doorway.

Portrait of Dr Frances O. Kelsey,
3 May 1965.

Advertisement for iodised salt.
Early twenty-first century.

the College's governing committee record that "A Scheme of a Charitable Design was offer'd to this Committee, from ye Committee of Medecines, & Read," after which the "Scheme" was "Refer'd" to the College's officers. To avoid the arguments that had scuttled earlier plans for a dispensary and in the spirit of "it is better to ask forgiveness than permission", in November 1696 the Committee of Medicines simply took up a private subscription to fund a dispensary and went ahead without putting it to the College for a vote.

The success of this private plan gave the Committee of Medicines courage on 21 December 1696 to propose putting a subscription to the general membership. They had already found a location for the Dispensary and arranged the terms, leaving the general membership with little traction for objections. In April 1698, the RCP opened the Dispensary's doors at their headquarters in Warwick Lane. Cue trumpet fanfare! The poor were saved! Previously, as one anonymous partisan put it, illness was the "ruine of the Poor, who beside the pain and dread event of the Disease, are under the fear of spending their whole Substance, in one sickness, and being absolutely undone". The Dispensary, however, "is the greatest Relief to the Poor, who have the best advice, and the best Physick at a very small Expence".

That last phrase – "at a very small Expence" – reveals that unlike, say, the Edinburgh College's Dispensary, the London College's Dispensary was not designed to provide access *for* the poor; it was designed to provide access *to* the poor. And "the poor", solvent or insolvent, were a vast market. Physicians also encouraged their middle-class patients to buy their medications from the Dispensary. The unnamed pro-Dispensary narrator of *The Present State of Physick & Surgery in London* urged his acquaintance to purchase them there. The supposedly charitable institution not only preserved but in fact promoted medication as a commodity. It also violated the College's Charter, which permitted physicians to inspect apothecaries' stores but not to make their own medication, as well as the apothecaries' Charter, which granted only apothecaries the right to make and purvey medication. The RCP knew it; committee minutes reveal that the Committee of Medicines reviewed the apothecaries' Charter in 1695. Dress up the Dispensary in benevolence and spritz it with the perfume of sanctity, however, and that attempt to seize part of the medication market – a very, very bad action – appears to be a very, very good action. Claiming benevolence covered the reality of the Dispensary and offered a deceptive justification.

The ensuing fight over the Dispensary not only pitted apothecaries against physicians but also pitted physicians who opposed infringing on the apothecaries' Charter against those who embraced it. Some opponents argued that each profession should do what it was trained to do, and that the different practitioners could and should work together for everyone's benefit (even the patients). As Mr Trueman of *Physick Lies A-Bleeding* states, "I am for employing every Man in his own way, the Doctor for Advice, the Apothecary for Medicines, and the Surgeon for Wounds, et cetera." Others were infuriated by the financial implications. William Salmon pointed out in 1696 that the goal of the physicians was "in plain English, to Monopolize all the Practise of Physick into their own hands". Sounding very much as if he had been reading John Locke's *Second Treatise on Government* (1689),* Salmon argued that

---

* "And reason, which is that law, teaches anyone who takes the trouble to consult it, that because we are all equal and independent, no-one ought to harm anyone else in his life, health, liberty, or possessions." John Locke, *Second Treatise on Government*, 2.6.

"a Monopoly [is] wonderfully prejudicial to the Lives, Liberties, Estates and Properties of the good People of England", especially the impoverished: "I hope the Mercies of God are such to the poor People of this City, that he will never permit" it. The battle over the Dispensary exemplified the thinking and actions that justified refusing medication to those who could not pay for it, regardless of the outcome. Although the combatants realized that there was a genuine problem of access to medication, they did not work to solve it.

The fighting was fierce; at one point, a brawl actually broke out at the RCP headquarters in Warwick Lane. Admittedly, physicians appear to have been a pugnacious lot at this time. Dr John Woodward and Dr Richard Mead once drew swords over an argument about the efficacy of a medication. When Woodward slipped and fell, Mead ended the fight by sneering, "Take your life!" To which Woodward retorted, "Anything but your physic!" A fist fight between two physicians at Child's Coffee House in 1734 made the news. As *The Universal Spectator* lamented, "Where shall we seek for Unity and Love / When Brother *Doctors* civil Discord move?" Fortunately, although less sensationally, both Dispensarians and anti-Dispensarians fought primarily in print, using the idea of caring for the poor to nurture the idea of medication as a commodity and thus to hide their prioritizing profit over access.

The most significant of those works was Samuel Garth's poem "The Dispensary", from which the title of this chapter is drawn. By the time Garth was writing those letters to Hans Sloane about "needy sluts", he had become wealthy and famous because of his medical practice and literary talents. His patients included dukes, duchesses, lords, and ladies, as well as Queen Anne's husband, Prince George of Denmark, and her successor,

George I. Garth's good friends among the literati included Joseph Addison, Richard Steele, William Congreve, Lady Mary Wortley Montagu (who brought inoculation from the Ottoman Empire), and Alexander Pope. He was considered the literary heir to the acclaimed playwright and former poet laureate John Dryden, a reputation encouraged when he arranged Dryden's funeral procession and burial at Westminster Abbey in 1701. Garth's own death in 1719 was so notable that newspapers reported it.

The timing of Garth's arrival on the scene and the resumption of the dispensary project by the RCP is suggestive. Garth did not receive his medical degree until comparatively late in life: in 1691, when he was thirty or thirty-one. After that, his medical career took off. He was admitted as a licentiate to the RCP (a step below Fellow) in 1692 and as a Fellow in 1693. Whether before or after he was made a Fellow, Garth and Sloane must have discovered an ally in the other; the dates suggest that Garth's admission to the RCP was based in part on his willingness to support Sloane in the debate over establishing a dispensary. In 1695, he joined the Committee of Medicines, uncoincidentally chaired by Sloane, and was among the first subscribers to the project. When the original proposed lease fell through, Garth was part of the group that took the new lease – on rooms at the RCP, thus giving the Dispensary a home and the RCP another form of income. He and Sloane remained close friends until Garth's death.

Garth used "The Dispensary" to frame the war as a battle between the benevolent and competent (the physicians) on one side and the venal and incompetent (the apothecaries) on the other. He cannily targeted a general audience and his poem proved hugely successful. To maintain control of the terms of the contest, he frequently revised and reprinted the poem to keep up with developments; it was reprinted with "corrections" three times in 1699 alone. Garth told his story in six cantos. In the first canto, the deity Sloth is awakened from his slumber at the RCP's hall in Warwick Lane by the sounds of people building the Dispensary. Since Sloth just wants to sleep, in Canto II he recruits the goddess Envy to help him end the project. She assumes the form of Colon, the most recent warden of Apothecaries Hall, and goes to alert Horoscope, another prominent member of the Worshipful Society of Apothecaries, to the Dispensary. Horoscope is so upset that he faints, but he is revived by his assistant, Squirt, who uses the "steam" of a urinal as smelling salts. In Canto III, Horoscope summons the apothecaries to their guildhall to decide on a response to the physicians (called the "Faculty"). The discussion is cut short when the laboratory under the Hall explodes. A disreputable crew of physicians and apothecaries meets in the next canto to conspire against the Faculty and destroy the Dispensary. Canto V recounts the battle between apothecaries and physicians, fought with medical instruments and bodily substances, until the goddess Health, also known as Hygeia, intervenes. In the final canto, she brings one of the physicians to the Elysian Fields to get the great physician William Harvey's instructions on how to resolve the battle. Harvey utters his lament that everyone cares about money and nobody cares about learning, and tells them how to resolve the conflict.

In Garth's representation, there is nothing inherently problematic about the Dispensary. The apothecaries are incited to opposition by Envy, an emotion for which the physicians cannot be held responsible. Attributing their opponents' objections to a personal flaw is a very tidy trick for sweeping away the physicians' wilful violation of the apothecaries' and physicians' Charters. It also neatly places the blame for the conflict on the

apothecaries: if they were not so envious, everything would be
fine. At the same time, it is disingenuous of Garth to identify
the cause of the apothecaries' ill will as envy because the poem
does not in fact depict the apothecaries as envious. Instead, they
are deeply concerned about having their secrets exposed. At the
apothecaries' meeting in Canto III, Ascarides the Elder wails,
"Suppose th'unthinking Faculty unvail, / What we, thro' wiser
Conduct, wou'd conceal." They also resent losing the sole right
and power to wipe out humanity. As Colocynthus explains:

> we [are] the Friends o' Fates, / Who fill Church-yards, and
> who unpeople States, / Who baffle Nature, and dispose of
> Lives

(One wonders if Garth is turning on the apothecaries the
adage "The young physician fattens the churchyard".) And of
course, they are more interested in money than in anything else,
including their patients' lives. Horoscope, for instance, "knows
that to be rich is to be wise".

On the whole, Garth's anti-Dispensarians, apothecary and
physician alike, are appalling specimens. Colon, the former
Warden of the guild, is "shrill" and "in Morals loose". His brain
is "empty", while "Hourly his learn'd Impertinence affords
/ A barren Superfluity of Words". Horoscope "fancies that
a Thousand Pound supplies / The want of twenty Thousand
Qualities" and Squirt is inordinately concerned with urine.
There are "Two Brothers nam'd Ascarides" who "Both had
the Volubility of Tongue, / In Meaning faint, but in Opinion
strong." Querpo, an anti-Dispensarian physician is a "worth-
less Member of the Faculty". As for his colleague Carus, his
"Spirits stagnate", his blood is "chill" and "sluggish", and
his mind full of "lazy Fogs". He is, Garth writes, a "brainless

Wretch". Mirmillo, a physician who confabulates with the apothecaries in Canto IV, is positively lethal and proud of it. "Whilst others meanly ask'd whole Months to slay, / I oft dispatched the Patient in a Day," he brags. Garth also implies that Mirmillo is attracted to men – he "Seal'd the Engagement with a Kiss, / Which was return'd by th'Younger Askaris" – a serious accusation in a highly homophobic time.

In fact, Garth accuses both Dispensarians and anti-Dispensarians of greed. Their battle in Canto V only halts when "Apollo interpos'd in form of Fee". Like Jane Barker in her poem "On the Apothecary's", discussed in the previous chapter, Garth relied on the association of Apollo and the sun to accuse physicians of wanting those little gold discs a little too much. "Some members of the Faculty there are," Askarides the Elder notes, "Who Int'rest prudently to Oaths prefer." Elsewhere, the narrator describes a scene of:

grave Physicians at a Consult met; / About each Symptom how they Disagree, / But how unanimous in case of Fee

Physicians even believe that they have a "right to assassinate" when it comes to money.

All this – the meetings in guildhalls, the satiric descriptions of apothecaries and physicians, the battle fought with syringes and gallipots, the intervention of deities such as Envy and Hygeia, and so on – is a Trojan horse. Both sides are greedy, both sides chase after fees, and whoever has this motive, Garth indicates, is wrong. Nor does "The Dispensary" ultimately justify the Dispensary on the basis of apothecarial inferiority, a frequent claim by Dispensarians: Canto V's battle is a draw. But the poem also does not justify the Dispensary on the basis of serving the poor, either. They only appear as two passing allusions, one on the usefulness of workhouses, the other as fawning praise of William III's compassion.

Instead, the poem justifies the clinic in the name of the Scientific Revolution. Garth bookends the narrative with laments that physic has abandoned the New Science. It is no longer interested in discovery and knowledge for improving the treatment of illness. After the Great Fire of 1666, Garth points out that the College's hall in Warwick Lane was:

> Rais'd for a use as noble as its Frame;
> Nor did the learn'd Society decline
> The propagation of that great Design;
> In all her mazes Nature's Face their view'd,
> And as she disappear'd, they still pursu'd.

Garth is talking about the study of the body through anatomy, that is, using the methods and principles of the Scientific Revolution. Following his allusion to the famous anatomy theatre ("a use as noble as its Frame"), Garth spends a long time devotedly enumerating the most famous breakthroughs in physiology. Much research remains to be done, he points out, especially in the connection between the flesh and consciousness: "we wait the wondrous Cause to find, / How Body acts upon impassive Mind". Unfortunately, "we" will be waiting for a while: "But now those great Enquiries are no more" and "The drooping Sciences neglected pine". Physicians "ne'er rifle her [Nature's] mysterious Store, / But study Nature less, and Lucre more". As William Harvey puts it at the end of the poem, "what was once a Science, now's a Trade". The whole story is boxed in by a lament that the noble medical profession has abandoned the Scientific Revolution for commerce – "trade".

Garth is not bemoaning the replacement of knowledge with

coin. Like other sons of the Scientific Revolution, he sees profit (in this case, "Lucre") as part of scientific discovery. His explanation of "How Physick her lost Lustre may regain" collapses money and learning into one. The College's penury prevents it from contributing to the New Learning; the solution is to get money for that noble endeavour through the Dispensary. To have a dispensary, however, the Charter must be changed. Via Harvey, Garth urges his colleagues to apply to the king, William III, to rewrite their Charter because that will facilitate medical research. "To him", Harvey/Garth orders Celsus/the physicians:

you must your sickly state refer,
Your Charter claims him as your Visiter.
Your Wounds he'll close, and sov'reignly restore
Your Science to the height it had before.

For Garth, this is the public statement of the Dispensary's purpose: to make the RCP sufficiently solvent that they can do more New Science. Providing free medical care for the poor never enters into the poem. Not once.

Garth was an enthusiastic adherent of the Scientific Revolution. That is not to say that he did experiments or published medical studies or any of the expected things from a revolutionary, because he did not, although Hans Sloane got him made a Fellow of the Royal Society anyway. Rather, he fully ascribed to its principles and methodology, and believed in its possibilities. Even after his death, Garth was associated with the scientific method. An anonymous tract from 1744 used the poet–physician in a satire of *Siris*, a posthumous work by George Berkeley, Bishop of Cloyne that endorsed tar water as a panacea. In *Siris in the Shades*, the ghosts (or "shades") of two medication purveyors in Hades ask Samuel Garth to judge the validity of each man's claim to have sold an excellent medication. Their argument is really an attack on Bishop Berkeley's tar water. Garth's ghostly opponents say things such as: "I do not pretend to understand Logic, for my Part; but I look upon *Tar-water* to be good in Fevers, because the Bishop of *Cloyne* has said so. ---And this has more Weight with me, than all your Arguments put together," following scholasticism's emphasis on authority over experiment. Having read *Siris*, Garth's shade concludes that he does not rate it very highly: "For, though I have read the Book, such a studied Want of Perspicuity, and such a methodical Want of Method, seems to run through the Whole that, I confess, I am not much the wiser for it." To which the second contestant offers the kind of retort that would strike horror into the breast of any scientific revolutionary: "Well, Sir, you may reason as long and as learnedly as you please, upon the Folly of trusting to Universal Medicines and the Retailers of them; but depend upon it, that we *Empirics* shall still have the Multitude on our Side." Garth sounds very much like his "Dispensary" self when he replies: "I grant it, Sir; and for this plain Reason, because it is much easier for a Man to surrender himself up to the Guidance of any Pretender to Medicine, than to be at the Pains of examining his Pretensions by the Light of his own Sense and Understanding."

There were those who saw access to medical care as a problem requiring a systemic change, who placed the Scientific Revolution's values and methods at the heart of a new vision for a medication system. In 1689, Hugh Chamberlen also proposed a method for expanding access to medical treatment designed to promote the New Science. Chamberlen belonged to the famous family of Chamberlens who invented the forceps and kept it a secret for nearly a century (*c.* 1600 – *c.* 1725) to command high obstetrical fees. The family exemplifies the interest in tools and technology, the privatization of knowledge, and the use of knowledge for personal gain baked into the Scientific Revolution. Chamberlen's idea for what he called a "more compleat practical Constitution of Physic" was published as *A Proposal for the better Securing of Health. Humbly offered to the Consideration of the Honourable Houses of Parliament*. It looks like an early version of the UK's National Health Service. In his view, too few people had access to any part of professional medical care, let alone the full range from diagnosis to cure: "as Physick is now managed", he wrote, "not only the very Poor, but meaner sort of Tradesmen and their Families, Servants, and Misers, deter'd by Physitians Fees, and Apothecaries Bills, have little or no Benefit by Physic". In his system, everyone would receive treatment: "all Sick, as well Poor as Rich, shall be Advised and Visited, when needful, by Approved, Skillful Phisicians, and Surgeons; and furnished with necessary Medicines in all Diseases". It would be paid for with a hearth tax graduated to each family's income, with additional charges for treating "the Pox", for attending labour and delivery, and for operations for "the stone". By "each House", Chamberlen meant "every Individual Person of the Family, as well the Lodger and Servant; as Master, Mistress, and Children". "It's proposed to serve all the Families," he explained. "Rich and Poor, Little and Great, within the City and parts adjacent, much better and cheaper than at present, with Visits, Advice, Medicine, and Surgery."

At first, Chamberlen's public health system appears to share a great deal with domestic medicine. Everyone can get good, reliable treatment regardless of who they are. "Household" is defined broadly to include everyone regardless of status. Food and drink are as vital to preserving health as medical care is to restoring it. Chamberlen advocated revising or passing laws "against the Sale of unwholesome Flesh in the Markets", "that Wine not [be] Sophisticated", "that Bread may be well Baked" and "Beer well Brewed". (He also advocated for public sanitation: "the Houses and Streets well cleansed from Dirt and Filth".) The foundational perspective of his proposal, however, is very different from domestic medicine's. There are no women, for one thing, only male professionals. His plan is based on laws, not on an ethos that holds the culture together. Legislation protects food and drink, those important components to health. He wants to reduce or eliminate "Epidemical and Contagious Distempers" and reduce the number of deaths, but because "To preserve Health and save Lives, is always a Public Good, but more especially in time of War." In other words, we need to preserve public health so we can send strong men into our armed forces – "food for powder", as Shakespeare put it. Health and treatment serve a function: to expand the empire, to protect the nation, to ensure a sound economy. They are not rights or part of daily life.

Chamberlen had the advancement of science in mind, at least. One advantage to his plan, he wrote, was that "Physic and Chyrurgery will be extreamly [sic] improv'd, and in little time, by the multitude of Experiments, recording of Observations, and mutual candid Assistant of the Members, come near to a Demonstration." Expanding access to medicine and medication does more than take care of the vulnerable members of society. It expands the number of people available for experimentation, observation, and the demonstration of results. London and its environs would be the largest colony of laboratory mice at any time in history.

Hugh Chamberlen's plan was never enacted, and the RCP closed the Dispensary by 1726. Personal charity, including individual apothecaries, private organizations, and the parish continued to provide medication and medical care for the poor, solvent and insolvent alike. Despite all the discussion and the recognized need, the commodification of medication did not feature any real attempt to adjust the system to provide access for the poor. Discussing this topic, however, and the way it was discussed deflected attention from and normalized what was actually happening: profit was being elevated over life. As "A Chymist in the City" caustically noted:

> If these Gentlemen had not made so great a noise about the Poor, and shewn such great industry in inviting to buy their Panaceas, one might have believed something of their Generosity. But Mankind now adays [sic] are not so immoderately desirous to serve one another, much less those that have most need.

6

The Laboratory
on Cheesewell Street

It is the early eighteenth century, you are passionately dedicated to the Scientific Revolution, and the dead elephant that you purchased has been delivered to your front lawn. You wanted to know all about elephants, but now that the thing is right there in front of you, you are having second thoughts. Dissecting an elephant may not be a one-man job after all. Also, your wife is furious. If you are Hans Sloane and this is London, the solution is obvious. You need James Douglas, the premier anatomist of the day. Besides, he has dissected a flamingo; the man's unshakeable.* Off goes a gracious invitation to your colleague to share in your good fortune. Unfortunately, James Douglas is no fool and he has anticipated you, stopping by to view the rotting pachyderm when you are not home just in case you extended this very invitation. The elephant's "enormous

---

* Douglas published "The Natural History and Description of the Phoenicopterus or Flamingo; with two Views of the Head, and three of the Tongue, of that beautiful and uncommon Bird" in the Royal Society's journal, *Philosophical Transactions* vol. 29, no. 350 (31 December 1716).

bulk quite frightens me from medling with its dissection," he replies. Sorry.

Douglas was not just the best anatomist of the early eighteenth century. He was also a leading obstetrician and a profoundly curious, deeply committed contributor to the Scientific Revolution. As soon as he earned his medical degree from the University of Rheims in 1699, he joined the corps of physicians at the leading edge of the revolution. He returned to London to practise obstetrics with the famous Chamberlen family, whose interest in tools and technology for enhancing human ability had led to their famous, secret obstetrical instrument, the forceps. Douglas took stunningly detailed case notes and, whenever possible, performed post-mortems. He shared his expertise through papers at the Royal Society, publications, and correspondence with other researchers around Europe, as well as training and mentoring young physicians. He was one of the first to give anatomy lectures out of his home to educate the paying public. The Company of Barber-Surgeons awarded him the Gale Osteology Lectureship for 1712 and the Arris Muscular Lectureship in 1716 to teach the apprentices. As Helen Brock, one of his twentieth-century biographers, understatedly put it, Douglas was a "worthy inheritor of the new approach to science and medicine, all his work being characterized by careful observation and the testing of traditional beliefs against new discoveries". He was the very model of a scientific revolutionary.

When he became interested in medication, he sought an expert in chemistry for help. Chemistry was in the process of evolving from alchemy, roughly defined as the belief and practice that every substance in the cosmos could be manipulated until it yielded the pure element or elements at its heart. Alchemy gave the emerging science the ideas that substances can be reduced to their component parts and that breaking down and recombining substances can yield an improved product. "Chymistry, is an art that doth both teach and inable us (for our exceeding good and benefit) seperate [*sic*] Purity from Impurity; Exalt and advance what God and Nature hath given us, to a farther and higher Perfection than we receive it indewed with," Richard Fletcher wrote in 1676. With any kind of substance, he explained, once its "essential purities are separated" from it, those pure substances "make it plain enough by their powerful effects, that it is to this state they ought to be reduced, before they work with Efficacy". Alchemy also provided chemistry with equipment and a set of techniques for transforming substances: boil, distil, pulverize, dry, burn, melt, and so on. The Scientific Revolution provided chemistry with a methodology – the scientific method – and a way of thinking.

Before this chapter goes any further, two interrelated but non-synonymous terms would benefit from definitions: chemical medicine and medical chemistry. "Chemical medicine" has several meanings. The term applies to medication made through chemical processes. That means transformations – many transformations – using fire and water and so forth. Consistent with the belief inherited from alchemy that everything could be purified and the pure version is always better, a compound or an ingredient was not just boiled once; it was boiled until it left only a residue, reconstituted, and boiled again three, four, perhaps nine or ten times. "Chemical medicine" also applies to Paracelsan medicine, which relied on metals and minerals. Paracelsus rejected theories of health and disease inherited from Hippocrates and Galen because those theories often clashed with evidence of the senses. His revolutionary and therefore supposedly more correct theory contended that malfunctioning organs caused illness, and medication's purpose was to poison or drive out the substance that was making the organ malfunction. A number of people spotted the problem with deliberately poisoning the patient to make them well: dosage was everything, and dosage was extremely hard to judge. A little too much lead and your patient will be beyond curing. Nevertheless, both aspects of "chemical medicine" caught on: medication created through certain ("chemical") processes that often used metals and minerals, and the school of thought that endorsed this kind of medication.

"Medical chemistry" is chemistry used in the service of medicine, especially to improve medication. Just as Paracelsus's ideas attracted many people dissatisfied with the ancient models of medicine, the possibility of repurposing alchemical methods to improve medical care attracted adherents who were looking for new types of medications. Medical chemistry's primary quest during the seventeenth and eighteenth centuries was to identify the curative component of any medicinal material. That goal is not the same as creating or discovering a medicinal substance. Take aspirin, for example. It comes from willow bark, which was known as far back as Sumer as an analgesic, antipyretic, and anti-inflammatory. Between 1826 and 1828, several individuals and groups used chemistry to extract salicin and identify it as willow bark's medicinal essence. In other words, chemists isolated a substance that everyone knew must be there from a plant that everyone knew about. It took another decade for chemistry to do anything more. Chemists transformed salicin into salicylic acid in 1838, turned it into acetylsalicylic acid in 1853, and finally made it into a powder patented and sold by the Bayer company in 1899. Chemistry did indeed contribute something: the transformation of salicin into a more effective compound. Just not during the Scientific Revolution.

In the seventeenth and eighteenth centuries, however, proponents of the Scientific Revolution thrilled to chemistry's possibilities. They believed that it could improve the phar-macopoeia, creating more effective, less dangerous drugs for sufferers. It could create a class of drugs based on transformed substances that women could not reproduce. It could establish all-male spaces and it could make them seem more impor-tant than female-associated spaces. Last but not least, because chemistry was new and a product of the Scientific Revolution, its improved medication would promote science and vice versa.

Initially, the home provided everything the chemist needed, including a woman (unless she was also the chemist) with experience and knowledge about the tools, technology, and processes. Many households also had a stillhouse or stillroom. Initially used to make simples – that is, to process herbaceous materials like roots, bark, leaves, flowers, and seeds into liquids with medicinal qualities – by the eighteenth century the space was used to make all household liquids, such as cordials, medic-inal waters, wines, and perfumes such as rose water. Stillhouses came with stills of course, as well as plenty of flat surfaces, all kinds of nifty equipment, and furnaces. The inventory of Charles Allington's kitchen in 1731 included:

> four brass porridge potts, two copper coffee potts, one chocolate D., iron grate, shovel, tongs, poker, fender, two tables, one brass warming pan, one pewter cullender, two pewter quart potts, one two quart pott, two pints eight blood porringers, dresser and shelves, one large bell mettle mortar, one small brass mortar and pestle, six chairs, a halfe quarterne pint and halfe quarterne measures of pewter, brass scales and weights…

The "Brewhouse" had "one large copper and iron grate, to the same one large mash pott and all other brewing vessalls".

Standard materials for a kitchen of the time.

There were a number of problems with using the kitchen to perform chemistry experiments, however. Kitchens and still-houses were constantly in use, so anyone wanting to do more than tinker with a specific recipe (think Sir Kenelm Digby when he was starting out) had to create a separate place with the same capabilities. Some chemists did not have a well-equipped kitchen, especially if they were bachelors. Besides, chemistry is smelly, not to mention occasionally explosive. It is best done in a place that cannot burn down. Early chemists turned to apothecaries, who had the necessary equipment and space, and who (especially in Oxford) allowed meetings or demonstrations in their shops after hours. Individual apothecaries began to imitate the Worshipful Society of Apothecaries of London and to build their own laboratories to provide medication in bulk.

Eventually, borrowing apothecaries' shops proved impracticable, and the laboratory was born. Peter Stahl, for example, needed his own laboratory in Oxford to help him teach chemistry. Isaac Newton had a laboratory; so did Samuel Pepys, now more famous for keeping a diary. Robert Boyle, once called "the father of modern chemistry", had laboratories at different residences during his life. Charles II had two laboratories: one at the palace of Whitehall, the other at St James's Palace. A chemist, like a housewife or a cook, might employ one or more assistants, who either helped the chemist or did the work under their direction. Advertisements for private "laboratory operators" began to appear in late seventeenth-century newspapers, such as the one in the *Collection for Improvement of Husbandry and Trade* in 1695: "One that has practiced Chimistry some Years, desires to serve a Chymist, or some Gentleman, that for his Diversion keeps a Laboratory." Laboratory operators could put such a position to good use. Like George Hartman with Sir Kenelm Digby's laboratory notebooks, John Headrich, "formerly Operator to Dr. Richard Russell", used his previous position to tout his book *Arcana Philosophica; or, Chymical Secrets* (1697).

The laboratory had some bad press to overcome initially. Sceptics of the New Science latched onto experiments as signs of foolishness or credulity. Thomas Shadwell's play *The Virtuoso* (1676) and Aphra Behn's *The Emperor of the Moon* (1687) very successfully satirized experimental science with protagonists who obsessed over pointless experiments and missed everything going on literally under their own noses. The silliest things in the plays happened in the laboratories, of course, as when Shadwell's Sir Nicholas Gimcrack tries to learn to swim by lying atop a table and flailing his limbs. In *The Emperor of the Moon*, after discovering that he was tricked into believing that two suitors of his daughter and niece are the Emperor and Prince of the Moon, Behn's Doctor Baliardo exclaims:

Burn all my Books, and let my Study Blaze, / Burn all to Ashes, and be sure the Wind / Scatter the vile Contagious Monstrous Lyes.

By the first decade of the next century, however, laboratories had lost a great deal of their mystique. When Susanna Centlivre satirized the New Science in her play *The Basset Table* (1705), she attacked it for using incomprehensible language and excluding women. Centlivre's female protagonist, Valeria, is an eager and intelligent observer and experimenter who keeps a laboratory in her home. Her suitor, Ensign Lovely (it is the eighteenth century, stage characters had names like that), admires both her and her experiments. "Oh you Charm me with these Discoveries," he says as an aside. Valeria's father, Sir Richard Plainman, is the villain of the piece, throwing her laboratory equipment out the window and forcing her to marry the man of his choosing. "You and your Will may Philosophize as long as you please, Mistress, but your Body shall be taught another Doctrine, – it shall be so," he sneers. "Your Mind, and your Soul, quotha! Why, what a Pox has my Estate to do with them?" Of course Valeria winds up with Ensign Lovely, who she loves as much as he loves her. As Centlivre's successful play indicates, by 1705 the laboratory is not the cultural oddity; a woman working in it is.

Medical chemistry was also established by then in its effort to identify the mechanisms of curative substances. When James Douglas, that model of a scientific revolutionary, wanted to know in 1723 what made medication work and whether it could be improved, he hired "Mr Durham's Laboratory in Cheesewell Street" to analyse "Chymical p.tions made by my order". The laboratory is long gone, but Douglas's "Cheesewell Street" was "Cheswell Street" on contemporary maps and "Chiswell Street" on maps today. Amazingly, however, the laboratory notebook survives. It does not look like much, just a slim, leatherbound book, but it contains the records of ten years of chemical experiments performed in Mr Durham's laboratory. True to the New Science's aspiration to objectivity, Durham kept meticulous records in a detached tone. With considerable *sangfroid*, a note at the bottom of one report warns future experimenters that "the Crucible ought to be pretty large, and take care that no charcoal get into the Crucible otherwise the matter will flame most furiously and fly out to some distance. this [*sic*] was our fate and perhaps that might have done harm to the Tincture." Sometimes, however, Durham proved human, as when he effused over a "glorious colour". The chemists of the day might not have known what they were doing or what they had made, but they could still feel wonder and delight.

Douglas first requested an analysis of wormwood, which makes good sense. It was used to treat a wide variety of ailments. According to Elizabeth Blackwell's *A Curious Herbal* (1737), the "Leaves & Tops" of wormwood "purge Melancholy Humour, provoke Urine, restore an Appetite that is lost by Drinking", and "are good against the Disorders of ye Stomach, vomiting and Surfeits: they strengthen the Viscera, kill Worms, & are of service in Dropsies, Jaundice, tertian & quartan Agues". Dr George Cheyne used wormwood to treat Prudence Wise's husband's "not intermitting" fever, and Prudence's own recipe for "Lady Nottingham's Oyl of Charity" for "scalds and burns" required it. Dorothy Rousby used wormwood in medicines for scurvy, rickets, and general aches and pains. The London apothecary John Doody treated digestion problems with it. Who would not wonder what made wormwood effective for treating so many different conditions?

So at Douglas's order, Durham's laboratory attempted the task:

We put 20 pounds of Wormwood into a large Copper still with a sufficient quantity of Water, fixing and luteing to it a large Copper head with a Leaden pipe to which fixes into the Worm, then setting under it a Seperatory [*sic*] glass vessel filled with cold Water, we put Under it a pretty strong fire once it begins to Work, and then we kept such a degree of heat as made it come over in a pretty full stream, the oyl that came over first was somewhat clear or yellowish but soon after it became black and thick, the Operation was begun about 11 ockloak [o' clock] and finished at 5 in the afternoon.

The Oyl was easily seperated from the Water in a glass funnel the oyl being higher than the water by allowing ye Water to run out we catch'd the oyl in the neck of the funnel.

We had of a black thick oyl 1 ounce.

The water that was of the still was of a fine red tincture and excessively bitter, so that I believe we might have got from it an extract and probably a fixed salt.

After six hours and after someone – probably the apprentice – tasted the mysterious red water that was left in the still, Mr Durham had no idea what they had made. Page after page of the notebook is the same: the laboratory operator describes what remained at the end of the procedure, but he could not identify it or say what it could do, if anything. One report ends with: "We put the first lotion of the Antimon. Diaphor: into an earthen pan in the sand heat and in 2 days time evaporated it to a dryness and got ʒiÿ of a Salt." Salt of what? What is it good for? No idea. The rest of the page is blank. Analysing oil of fennel seeds, the recorder concludes, "the Oyl was seperated by a Glass funnel and then passed thro a little cotton to free it from a whitish froathy substance that comes over in distilling all seeds we got ʒiss of it," "it" being the unknown "whitish froathy substance". All they can be certain of is that it appears every time they distil seeds.

The laboratory records of others follow the same form: detailed narrative of procedure, description of resulting substance, non-conclusion. When Daniel Coxe tested tobacco in his laboratory, "The oyle being separated from the Spirit a young man that was present tooke the 6th or 8th part of a drop on the tip of his tongue whose unpleasant tast [*sic*] though it made him immediately spit forth most thereof." Then the handy young man (probably another apprentice) got dizzy for two or three minutes and nearly fell down, after which the tobacco oil "did almost prove Emetick bellicating his Stomack as he affirmed, in a very troublesome manner; the vertigo and rachings [*sic*] to vomitt continued almost a ¼ of an hour". "I was deterred by his example from prosecuting this Experiment by triall on my selfe or on other persons," Coxe wrote without irony. He was nothing if not resourceful, however, and:

> having procured a lusty Cat, I put one drop of this oile into her throat, who immediately thereupon (as neer as I could guess w^{in} a minute or two) was taken with a staggering & fell down on her back where shee was grievously afflicted with Convulsions; I willing to Save the Catts – life [*sic*] reared her on her legs & endeavoured to keep her in motion.

At first the cat was stupefied, then she revived and began running around, and finally she took "a great leap fell down stone dead, without sense, pulse, or motion [*illeg*] the Trunk of her body and her limbs going as pliant as a rag". It has to be said: curiosity killed the cat. It also demanded a post-mortem. "As soon as I was satisfied that the cat was really dead, my Curiosity prompted me to open the body, and enquire what alterations the poison had made therin," Coxe adds. "As soon as I opened the skin my nose was saluted with ungrateful steams perfectly resembling the smell of the oyle." Otherwise, he concludes, there was nothing of note. Tobacco remained part of materia medica, although one hopes that no one attempted to prescribe oil of tobacco after that.

It did not go unnoticed that nobody was learning much. "How little do we really know of the action of other bodies on ours?" demanded George Martine in 1740, "And yet how readily do we at the very first pronounce all purges to act indiscriminately, differing only in their various stimulating forces?" But proponents of the New Learning were undaunted. In fact, the enthusiasm for chemical medicine drove a shift towards Paracelsan medicine. Metals, elements, or compounds with one or both comprise the majority of the experiments ordered by Douglas of the Chiswell Street laboratory. All but two of the experiments between number thirty-six through seventy-nine, the final one, analyse forms of sulphur, tartar, and different kinds of salts. Over the years Douglas's prescriptions became more chemical. On 24 August 1711, for instance, Douglas ordered a laxative for a patient:

Rhubarb. ℥iÿ
Curr: ℞vi ℨi. M

It was an electuary – a sweet, medicinal paste – of powdered rhubarb mixed into a kind of currant and potassium jam. A decade later, Douglas's prescriptions looked like this:

Elect: Antiepilept.
R. Ead: valer. sylv. ante
quam Caulem edatcolect
pulv. ℥iÿ. Cinab: Antim:
ℨi. Cinab: Antim: ℨi.
syr. Caryoch: 9:f. f:
El:

This electuary had a herbal component – "valer. sylv", or wild valerian (which Douglas wrote a paper about for the Royal Society) – but it also used a lot of cinnabar and antimony. By the end of his life, Douglas's personal *Catalogus Pharmacorum* included a "Catalogus Chymicorum" section as large as the catalogue of organics.

Other physicians' records confirm this shift. Between 1706 and 1709, the medications given to Sir William Millman's family were more often organic than Paracelsan, but not exclusively so. Medicaments included "prepared" pearl, Conserve of Athermes, and "Red Led plaister" as well as "Peny royall water", "Strong Cinnamon and Plantane water", "Syr[up] of Cowslips", and everyone's favourite, "Frog Spawn water". Eventually, some physicians' prescriptions only used metals and minerals. To make "the Pill", Dr Huxham instructs:

> Take Quicksilver make an Amalgama with p: ae. Regulus Antim. Mastielis [?] & pure Silver adding a proportionable Quantity of Sal. Ammoniac. Distill off the Mercury by a Retort in a glass Receiver then with the Quicksilver make a fresh Amalgama with the same Ingredients. Distill again & repeat the Operation nine or ten times then dissolve the Mercury in Spirits of Nitre put it into a Glass Retort & distill to a dryness calcine the Caput Mortuum 'till it become of a Gold Colour, burn Spirits of Wine on it & keep it for Use.

There is a lot of metal in the Pill. A lot. True to chemistry's alchemical origins, the mercury, antimony, and silver are combined, distilled, and recombined many times with sal ammoniac, which is a compound of ammonia and chloride, and spirits of nitre, a combination of water and nitric acid. Huxham's recipe notes directly link medical and chemical knowledge. "It is impossible for any one that does not attend the Process to Specify the precise dose," he wrote, "because the Medecines [*sic*] will be stronger or weaker as the Process is conducted. In general 30 Gr. Of the Antimonial Powder, & one grain of the Mercurial Powder is a moderate Dose, tho sometimes more or less is required."

Chemical medication had impressive commercial value. Unlike organic medication, it seemed obvious that it could not be made at home. The ordinary housewife was unlikely to own all the necessary equipment for making chemical medication, nor was she likely to be able to get hold of the materials required. Mercury the plant could be grown in a garden; mercury the metal could not. Her location (Was she urban or rural? What grew in her area? What was sold nearby?), weather (Was the spring too wet or too dry for certain herbs or flowers to grow?), and season (not a lot of fresh sage in January) also affected stock and therefore medical capability. Paracelsan medication did not have to wait for chamomile to grow or borage to flower. Its practitioners did not have to depend on the overseas trade for nutmeg or tamarind, either. From a sales point of view, chemical medication was new, mysterious, unavailable at home, and consequently eminently lucrative.

In the mid-1660s, practitioners led by Thomas O'Dowde (Mary Trye's father, last seen in Chapter 4) attempted to found a Society of Chemical Physicians and petitioned for a charter as a College of Chemical Physicians. Had they succeeded, they would have competed directly with both apothecaries and university physicians for patients and money (what "The Dispensary" mockingly deified as "Fee"). Sceptics of the self-styled "chemical physicians" viewed them as economic opportunists rather than true devotees of the healing arts, a view reinforced by the fact that many of the signatories were self-taught, untrained, and "illiterate", that is, they could not read Latin. Despite what appears to have been genuine belief in chemical medicine for some, the petition to form a new society failed and no economic or professional wall was built around chemical medicine.

The even bigger, game-changing issue was normalizing chemical medicine, regardless of why or whether it worked. As ever, without demand, supply is just the stuff lying around the house. Proponents argued for chemical medicine's place in education. As early as 1651, people like Noah Biggs were urging universities to add chemistry to their curricula. The pharmaceutical forged in a laboratory, Biggs proclaimed fervently, was "true medicine". Universities hired professors or endowed chairs to teach the newest methods and findings in physic, anatomy, and chemistry. At the same time, popularizers began working on the public. Readers were inundated with books such as George Starkey's *Natures Explication and Helmont's Vindication. Or A short and sure way to a long and sound life: being a necessary and full apology for chymical medicaments* (1657) and Peter Shaw's *Philosophical Principles of Universal Chemistry* (1730). George Wilson's *A Compleat Course of Chymistry* (1703) even included explanations of how different kinds of furnaces worked. It sold out three print runs in its first year. The great chemist Nicolas Lémery's *Course of Chymistry* was published in French in 1675; it appeared in English in 1677 and was reprinted in 1680, 1686, and 1698. A posthumous edition came out in 1720.

Chemists offered public lectures in chemistry, often in coffee houses, and sometimes expanded their impact by publishing those lectures in a collected set. (Coffee houses were busy places in the seventeenth and eighteenth centuries. Patrons could get a newspaper, conduct business, see patients, and hear a lecture on chemistry or Newtonian physics, for instance.) Distinguished physician John Freind's lectures on chemistry became available in 1712; John Quincy's lectures were so popular that, like Lémery's *Course of Chymistry*, they were published post-humously in 1723. Following the publication of his book *Philosophical Principles* in 1730, Peter Shaw published his lectures as *Chemical Lectures* in 1734. By the end of the 1730s, the public was hooked. When a "Laboratory [was] fitted up at St. James's [Palace] for the Use of his Royal Highness the Duke, who is going thro' [*sic*] a Course of Chymistry" in 1738, it made the *London Evening Post*.

Women could learn about chemistry, but only to some degree, and they were not encouraged to think of themselves as chemists. From one point of view, it was logical for women to learn chemistry and chemical medicine: alchemy and chemistry obviously overlapped with women's domestic work. The marvellous transformation of materials happened every day in kitchens where beer was brewed, milk became butter, meat was browned, and bread rose – never mind the medications that were made. Then there was the principle of it. As the author of *The Summary of Chemistry* explained in 1712, "It may indeed be a Crime in Politicks, but never in Good Manners to aim at a General Advantage; and hereby Ladies (if they please) are not exempted from attaining the Intelligence and Knowledge of Arts." Only a boor would exclude ladies from the "Knowledge of Arts". He also makes it a matter of national interest: knowledge of the sciences is "a Happiness which renders French Ladies particularly famous, where Sciences and Philosophy are made Natural to them, by being written in their Native Language; and sure it is but a barbarous part to debar that to others which we so much seek after our Selves".

It is not a matter of enabling women to set up their own laboratories and experiment with sulphur and mercury, however. Normalizing chemical medicine meant delegitimizing the old and replacing it with the new. The rhetoric around cooking and chemistry made divisions between them. "Consider our Bread, our Beer, Wine, Meat etc. Or whatever can render our lives happy or satisfactory: And you will find it in one degree or other to pass under the hand of Chymistry, and its various Operations, or Preparations," Richard Fletcher wrote. So far, so good. But he credits Nature, not women, with the use of chemistry to produce food, drink, and medicine. Women's tasks – making food, "Extract[ing] any thing from her Physick", and so on – he assigns to Nature. Fletcher also depicts Nature as "Diseased" and weak, unable to complete her domestic round without the help of male physicians. "For of Necessity", Fletcher concludes, "either the Physician or Nature must offici-ate, or act as Chymist." His formulation and language retain a female doer, but she is an abstraction rather than an actual female, and even then the female abstraction (Nature) cannot do her job without the assistance of a male physician.

The impact on women appears in their recipe books, a place where medication, cookery, and process converge. As a body of work, these texts show signs that women began to replace their own modes of thinking and measuring with those of male practitioners. No switch was flipped, no single transformation in recipe books occurred. Put together the dif-ferent kinds of change, however, and broader transformation appears. For one thing, increasing numbers of recipes from physicians or apothecaries started turning up in recipe books over time. It is a bit tricky counting contributions from apoth-ecaries because unlike physicians, apothecaries did not have a special title to distinguish them from ordinary people, and men did contribute to women's books. I have erred on the side of caution here and counted someone as an apothecary only when specifically identified as one.

Conservative estimates notwithstanding, the change in recipe sources was real. When Elizabeth Strachey began her recipe book in 1693, contributors included friends like Mrs Langton and family such as "My Mother" and "Aunt Clarke". Starting around 1727, the book began to include recipes from professionals. Prudence Wise's book, which spans roughly the same years as Elizabeth Strachey's, also follows that pattern. The first recipe from a physician does not appear until page seventy-seven, and then it is one that was in circulation among her friends. After that, physicians' names begin to appear frequently, some of them because the man himself contributed a recipe or two. "Mr Beaumont a Duguist [sic]" gave her "Doctor Meads prescription" for treating "an intermiting [sic] feavor", and a whole slew of physicians' recipes came to her from friends a little later: "A Receipt from Dr Ratcliff by Mrs Aislaby – for a Flux" and another "From Mrs Aislaby also", as well as "Doctor Palmers Electuary for the Stone or Gravell – from Mrs Greville December the 19th 1731", for instance. Some recipes seem to have come directly from the physician, as when she recorded "Doctor Mead Prescriptiosed [sic]" and "From Dr Cheney at Bath. February 1731." Sarah Draper's book includes recipes in different hands, including "For a Cough or Hoarseness" about forty pages into the book: "From 4 to 12 Drops of Balsam of Sulphur on moist Sugar now & then. As the Balsam is made of different Qualities by the Chymists, it might be necessary to ask them." There are several explanations for an increasing presence by professionals, but they all come back to the same point: over time, the family decided, consciously or not, to depend more on a male professional.

Another trend seems to have begun roughly around the 1720s: including medicinal recipes from print sources. Strachey started her recipe book in 1693; in the 1730s she added "ye following Receipts [...] out of *Salmons Practical Physick*" and "Dr Mead's Method & Meison for ye Bite of a mad Dog taken out of Farley's News Paper August ye 30 1735". Two recipes clipped from newspapers, one of them dated from 1747, were inserted at the end of the book. The owner of another recipe book slipped in a page from a printed book providing a "Method for Reviving Drowned Persons" (strip the body, lay it in front of a fire, and rub it all over with salt). In a number of cases, women treated a recipe from a physician the same way. Anne Selden copied into her own book the recipe for "Dr Lacocks Milk Water" given by "Mrs. Willoughby". Recipes given to her by physicians or apothecaries, however, she kept on the original paper pressed between two leaves. Another keeper of an eighteenth-century recipe book did the same, storing Dr Thompson's recipe "To keep the Body open" between the first two pages. In these cases, the recipes coming from outside the book owner's circle are treated as foreign objects. They are not recipes, they are prescriptions.

Recipe books also show that women began adopting chemical medicine's vocabulary. The recipe book is a genre; in earlier chapters I discussed how it has its own textual conventions, elements that all recipe books share. For example, recipe titles often specify what the recipe is for – "An approv'd Powder to prevent miscarriage", "For ye Gout", "An Electuary for the palpetations of ye Heart", "A Poultis for a sore Brest", "To make Wallnut Water for all malignant Diseases, or Poyson, or Surfett". Where the purpose is not in the title and it is not a common compound, such as "The Gray Ointment", the usage often comes at the end: "Approv'd for a Burn or a Scald". Instructions begin with a command, usually "Take"; provide the ingredients and their amounts as they go; and conclude with how to use the medication. "Let the Patient drink from half a Pint to three Pints every morning—according to the strength of the disorder," orders one book.

Some recipes provide dosages. Many of the recipes in one book from the early eighteenth century come with different dosages for children than adults, as well as for children of different ages. Abbreviations are rare and measurements are domestic: a handful, a "pennyworth", a pint, "as much as will lie on" a coin or the point of a knife, and so on. The further towards the middle of the eighteenth century the recipe books get, however, the more likely that professionals' terms will turn up. Ingredients are measured in drachms and scruples, the quantities sometimes written in longhand and sometimes with the symbol. One early book has apothecary symbols and their meanings at the top of the first page. Another recipe book refers to "Chemical oyls" for a set of ingredients that do not sound particularly chemical: "oyl of Lavender, Sweet Marjoram, Rosemary, Chymicall Oyl of Nutmeggs & oyl of Nutmeggs by Expression, & oyl of each half an ounce". Sarah Draper included antidotes for poisoning by different chemical substances – arsenic and oxalic acid are the first two – where there was extra space at the bottom of a page. Arsenic and oxalic acid are not the sort of thing lying around the ordinary eighteenth-century home, and they are toxic, if not fatal, for humans. Hopefully, she did not record antidotes because she had experience with poisoning.

The lexicographical invasion also takes more obtrusive forms. Prescription form begins to make its way into later recipe books, even if the recipe is all in English and does not use any code. After copying a recipe from Mrs Hyde "For a Strain" ("A penny worth of oil of Saint John wort, as much of oil of Exeter, as much of oil of earth worms" – that is the whole thing), the owner of an eighteenth-century recipe book recorded "For an Ague":

Take ½ a pint of Ale
a quart^r of a pint of Carduus water,
½ a pint of mint water,
Two pugils[?] of Sena,
Tamarrinds [*sic*] as much as a large nutmeg, boil all these together, over a gentle Fire till it be ½ a pint, & drink it an hour before ye fit comes.
If it be for a Child, ½ this, for a Man, or woman all of it.

When a prescription appears, it can be a startling intrusion. In one seventeenth-century recipe book, a recipe in full apothecary code for a "Strengthening Plaister" has been squeezed in at the bottom of a page. In fact, the juxtaposition of domestic and chemical can cause a bit of whiplash. On one page of a book at the RCP in London, the owner wrote this recipe:

To cure a sore Breast
Rx. Empl: de Panis cum ♂rio ₃
Somnj ferj theraic ₃ij
Take of ye Emplaisyer of Frogs wth Mercury two Ounces.
Opium ₃ii mix y^m & make a plaister for ye Breast.

On the facing page, she wrote this one:

An approv'd medicine to procure deliverance of a dead Child
Take 3 Dragon Roots & stamp y^m & divide & bind to ye hollow of ye Feet:

These recipes are both likely to fail, but they are very different approaches to medication. As in the top recipe, ingredients like mercury start turning up in women's recipes that are meant to be made. Elizabeth Smith, who began her book in 1700, used flour of brimstone as part of a treatment for asthma (brimstone is a type of sulphur; flour of it means that it has been turned into powder). In a later recipe book, a contributor has copied a recipe for "an ague" with measurements in ounces and drams (written out), as well as "Steel prepared with Sulphur half an Ounce". Such recipes indicate that women saw Paracelsan medicine – or just chemical medicine – as legitimate, equal to, and possibly better than the organic recipes that had been passed down for generations. Recipe books suggest that the idea was making its way into the cultural psyche, including into theirs.

Household accounts offer another view of these developments. Generally speaking, accounts reveal a lot about a family – consider the family that used up a half pint of gin three times between 14 and 23 March, for example – and they offer evidence that women turned increasingly to professional help and prepared medication. Allowance must be made for specifics, of course. Lady Doyly incurred an impressive set of charges for professionally made medication between August 1688 and May 1689, suggesting that she was unable or disinclined to rely on herself or her housekeeper for what appears to be innumerable ailments.

But accounts like Anne Brockman's, which span decades of married life, bear out this change. In the early days of her marriage, Brockman bought more ingredients for medication than she did prepared treatments. On 24 February 1700, she purchased eight ounces of liquorice, four ounces of anise seeds, and two pounds ^{of} runing [*sic*] Treacle, and on 17 April she bought them again: "2 lb ½ of Treacle" and a pound of liquorice. In May 1704, she bought carduus seeds and centaury, both used in herbal medication. She was still regularly buying liquorice, anise seeds, and centaury in 1721, but the amounts had increased considerably: in July alone, she purchased six bunches of centaury. Similarly, in 1701 Brockman bought "1 ½ ʒ of Syrop of popyes" and "3 drams of pouder of Corrall", and "10[?] ½ of lenitive Electuary" in 1704. She regularly purchased anise seed water. By 1721, however, "Hysterick water" was appearing frequently in the accounts. Perhaps the Hysterick water was for her – payments for nursing care that year, including a poignant payment to a nurse for helping to wash Mr Brockman, suggest that her husband was incapacitated and that she was running the household and managing the family's extended affairs – a very stressful situation.

Fortunately, the elevation of chemistry and chemical medicine did not establish Paracelsan medicine as "the way" to understand health, illness, and medication. Promoting chemical medicine changed perceptions of medication, however. Highly manipulated ingredients, such as those resulting from multiple transformative steps, came to be seen as the most effective ingredients. Because those manipulations, those transformations, could only be performed in a laboratory, the laboratory became the best site of medication manufacture. Laboratories were not domestic space (the very thought!), not kitchens, and therefore not female space; laboratories were separate from the home and were therefore male space. The residue of Paracelsan medicine's moment in the sun was the perception that medication is not made from natural or organic substances. Drugs for treating illness are made by men, in a laboratory, using special processes, and involving mysterious substances that are not – repeat, not – grown in a garden or anywhere else.

Katherine Jones, Lady Ranelagh, the most important woman in early chemistry, would have been outraged, and not only because she was a woman. Like James Douglas, she was a committed soldier in the Scientific Revolution, but she had a great deal more money and influence than he did. She used her wealth, social position, and influence to advance the New Science, especially to support chemistry. She herself was a chemist with her own laboratory at home in London in which she did her own quite serious and intelligent experiments. Ranelagh worked with and among other great scientific revolutionaries of her time such as Samuel Hartlib, Henry Oldenburg, the botanist John Beale, chemists Daniel Coxe and George Starkey (both of whom we have met in this chapter), physician and neuroscience pioneer Thomas Willis, and yes, Sir Kenelm Digby. Ranelagh was crucial to the brilliant career of her brother, Robert. She helped him establish several of his laboratories by sending equipment to him in the country, helping him settle at Oxford and set up his laboratory there, and allowing him to build a laboratory onto the back of her own home when he moved to London. It was quite a thing to do in the very fashionable neighbourhood of Pall Mall, but she also hired Robert Hooke as the architect, so it was probably quite elegant. For the rest of their lives, she and Robert occasionally performed experiments together, and they developed a medication for children with rickets that used copper.*

Lady Ranelagh brought together the new and the old. As a gentlewoman, she had been trained in domestic medicine, and she took her responsibility to provide treatments

---

* I have been unable to find evidence of its effectiveness.

and care for the sick and injured of her family and dependents very seriously, becoming renowned and respected for
her knowledge and skills. Collecting medical information in
the 1650s, Samuel Hartlib noted that she had told him that
"Walnut-honey (or the rindes of Walnut) is a most excellent
and soveraigne Remedy against a sore throat," adding that,
like many women with a household, "My Lady Ranalagh [*sic*]
hath store of it." She and Elizabeth Grey, Countess of Kent
discussed and exchanged medicinal recipes. She was a protec-
tive mother; her concern for her children sometimes meant
challenging diagnoses or treatments when they conflicted with
her own judgement and knowledge. She raised her daughters
to do the same, training them to treat their families while also
keeping up with medical developments. Ranelagh's extended
family sought her advice and often received medications from
her when ill. So did others to whom she was not related, including the royal family. She and Thomas Willis were among those
who attended the Duke of Kendal, the infant son of Charles
II's heir, James, Duke of York, when the baby was fatally ill,
and she was one of the few who read the autopsy report. She
also attended the little Duke's grandmother, Lady Clarendon,
in her final illness.

Overall, Ranelagh's medical practice drew on both domestic and chemical medication. She depended on her own skill
and knowledge but was willing to consult experts; she used
her own recipes as well as theirs. Indefatigable in pursuit
of good cures, she once exerted tremendous pressure on a
Dublin practitioner to give her his recipe for treating kidney
and bladder stones because her friend Samuel Hartlib suffered
from "the stone" so badly. Their correspondence indicates
that despite Monsieur Fontaine's resistance, she still extracted
almost everything from him. What mattered to her was

whether a medication worked, not what it was made of or where it came from. Practically and idealistically, Katherine Jones, Lady Ranelagh was both a person and a sign of the Scientific Revolution's possibilities.

She was neither the first nor the only woman with an interest in alchemy/chemistry and medicine. Her mother's generation (Elizabeth Grey and Alethea Talbot, for instance) had also been interested in alchemy and the first stirrings of chemistry, and women in Europe were already publishing works on chemistry, such as Isabella Cortese's *I Secreti Della Signora Isabella Cortese* (1595) and Marie Meurdrac's *La Chymie Charitable et Facile, en Faveur des Dames* (1656). Rather, Ranelagh is significant for proving that scientific principles could be reconciled with those of domestic medicine. On a practical level, she demonstrated that the pharmacopoeia could include any medication that worked, and that all medication should and could be tested for efficacy. Combining domestic and chemical medication was a logical step, and Katherine Jones, Lady Ranelagh showed how possible and effective that synthesis could have been. The crucial point is the "where" of medication. Despite the seeming compatibility of domestic and chemical medication, the two were firmly differentiated on the basis of where and how they were made. Even if you did not use metals, if you made your medication with chemistry's processes and, even better, did so in a laboratory, that medicine had legitimacy (efficacy is another matter).

Dr James Douglas and Katherine Jones, Lady Ranelagh shared a commitment to the Scientific Revolution and a deep belief that its methods and ideals could be used to improve humanity's condition. Both saw chemistry as a valuable tool for understanding the world and helping others. Instead of reconciling chemistry, medical chemistry, and domestic medicine,

however, the Scientific Revolution's acolytes went in an opposite direction, using chemistry and medical chemistry together
as an opportunity for profit and control – of medication, of the
market, of education. Most physicians and apothecaries, like
James Douglas, continued to depend more or less on organics
while focusing the culture's – and the paying public's – attention on synthetics. Chemical medicine's primary usefulness at
the time was as an opponent to domestic medicine, another way
of separating women and medication and rewriting cultural
norms and expectations. Medical chemistry nurtured aspira-
tions that it could not achieve; it did not improve medication
during the seventeenth and eighteenth centuries. Nevertheless,
by the middle of the eighteenth century, chemical medicine's
image as the "best of all possible methods and treatments"
loomed too large in the cultural consciousness to be dismantled
without concerted, communal effort.

7

The Doctoress's Cure
for the Stone

Bladder stones are painful. Really, really painful. They
are so agonizing that before anaesthesia and antisep-
tics, people willingly subjected themselves to surgery to
have them removed, and if voluntary organ surgery without
anaesthetic and with a high mortality rate does not indicate just
how unbearable it was, it is hard to imagine what would. After
seventeenth-century diarist Samuel Pepys survived his procedure on 26 March 1658, he hosted an annual dinner party to
celebrate the anniversary of the excruciating event. The stone
was on prominent display.

The seventeenth and eighteenth centuries called Pepys's
affliction simply "the stone", and it was appallingly common.
Lithotomy, the surgical procedure to remove stones, was first
developed centuries before Hippocrates, in the fourth century
BCE. By the middle of the seventeenth century, two primary
methods of lithotomy had been developed, the lesser (incision through the perineum) and the greater, or higher (incision
through the belly). Fortunately, the Scientific Revolution precipitated the development of better techniques: the "lateral"

method, developed by a French priest, and the "high" method, developed by John Douglas, brother of Dr James Douglas (discussed recently in Chapter 6). By the late eighteenth century, a medical student might receive several lectures on bladder stones and the appropriate surgical procedures. Alexander Gordon recorded in his notes that Alexander Munro, physician and professor of anatomy at Edinburgh University, described in his lecture different ways for operating on men and women. Dr Munro also passed around examples of bladder stones when he spoke on surgical interventions, warning his students that it was vital when making incisions into the bladder to be alert to bladder stones' variations in size and shape. Regardless of method, speed was vital. The great surgeon William Cheselden was renowned for being able to complete a lithotomy in just over one minute (once, in fifty-four seconds). Assistants were equally vital: someone had to hold down the patient. John Douglas used seven assistants in his operations.

First, however, the surgeon "searched" the patient. This he did by inserting an instrument called a staff up the patient's urethra and into the bladder. Its purpose was to confirm that there was at least one stone and its location – ideally, the surgeon might even get a sense of the size – so when the doctor went in, he knew exactly where he was going and what he was looking for (see above: speed). Doctors were more casual about searching than about operating. Edward Nourse, a Fellow of the Royal Society and a respected member of the barber-surgeons' guild, reported that he searched one "Mr Gardiner" at Childs Coffee House, which was indeed a coffee house. Mr Nourse did so "in the presence of several Physicians & Surgeons, who likewise felt for the stone". Happily for Mr Gardiner, although he had complained of bladder stones previously, he had none at this examination and was spared the knife.

Unsurprisingly, medicinal treatments were highly preferable. John Moyle, author of *Chirurgus marinus: or, the Sea-Chirurgion* (1702), instructed aspiring ship's surgeons to treat patients with medication and offered recipes for healing abraded ureters, easing pain, and dissolving stones. Among other recommendations, Stephen Blankaart's *The Physical Dictionary* listed goosegrass, marshmallow (a plant), and swallow grass for the stone. An abridged list in the 1702 *Pharmaceopoeia Londinensis* of plants, trees, and fruits useful for treating the stone included alehoof, turnsole, liverwort, ash bark, asparagus, burdock, balsam, pennyroyal, bayberry, lemon, beans, birch bark, broom, furze, spikenard, chickpeas, figs (dried or green), eglantine, eringo, galangal, goatsbeard, gooseberry, marshmallow, melilot, onions, pellitory of the wall, ragwort, saxifrage, smallage, swallowroot, tarragon, toadflax, and wild cherries. *The County Physician* (1701) recommended that its readers "Take a Nutmegg and cut it into four quarters, and steep it in good Sallad Oyl, 24 hours. Take a Radish Root, slice it and steep it in good white Wine, eat a quarter of the Nutmegg in the morning, and drink a little of the Wine above it, and fast two hours thereafter."

Personal recipe books were full of recipes for "the stone". One recipe combined "1 Gallon of Gascon Wine Ginger gallangall Cinamon Nuttmegs, grains, Cloves, Mace, illnesseeds fennel Seeds Caraway Seeds of Each 1 Ounce Sage red mint Rose leaves time Pelitory of Spane Rosemary wild time Chamomile Lavender each 1 handful" for a miracle drink that cured (among other afflictions) bad breath, infertility, head colds, and bladder stones. A slightly later recipe book recommended drinking up to three pints of a "strong decoction of black Currant leaves or in Winter, when the Leaves can't be produced, the small Buds peculiar to that Tree". Lady Sedley collected several recipes, among them "The Duke of Monmouth's Receipt for the Stone", "A Most Excellent Pill for the Stone or an ulcer in the kidneys or Bladder, or bloudy water. Approved by Able Physicians", and "Docter Jacob of Canterbury medicine for the Stone".

Naturally, printed recipe books also had them. *A Choice Manual* (1653), attributed to Elizabeth Grey, Countess of Kent, offered six: five under "A Remedy for a Fit of the Stone" and one from Sir John Digby. Readers could still find recipes for treating the stone almost a century later, testament to how often it afflicted people and how truly horrific the pain was. Charles Carter's *The Compleat City and Country Cook: or, Accomplish'd Housewife* (1732) did not encourage women to make and dispense medication, but it did advise women that birch wine treated "inward Diseases as accompany the Stone in the Bladder". Printed compilations of physicians' prescriptions, which began to appear around the turn of the eighteenth century, also offered many recipes. *Dr. Lower's and Several Other Eminent Physicians Receipts* (1701) listed seven for "stone" – the book actually offered nine. Other ailments might be permitted to vanish from the housewife's capabilities, but not kidney and bladder stones.

Desperate people will try anything, however, and ready-made nostrums abounded. There was the Tinctura Mirifica, which "infallibly cures the STONE and GRAVEL, whether in the KIDNEYS, URETERS or BLADDER, and also the Strangury, Stoppage of Urine, and all Heat, Pain, and Difficulty making Water" with just one dose and "in a Minute". And there was "The famous POWDER lately found out", which was advertised in the *Daily Courant* and "so much better than Tippin's Water (which hath had such a Name these several Years for that Distemper)". Anthony Daffy claimed that his Elixir Salutatis could cure kidney stones as well as gout, colic, diarrhoea, tuberculosis, and "ptisick". The Grand Specific cured kidney stones, as well as everything from incontinence to blockage. (It also reversed debilitation from "tedious or ill-manag'd Cures of the Venereal Disease, or from Self-Pollution, inordinate Coition, etc.") *Fog's Weekly Journal* in September 1737 tactlessly (or perhaps waggishly) placed advertisements for rival medicines, the Pulvis Salutis and Tipping's Cordial Liquor, next to each other on the page. According to its maker, the Pulvis Salutis could cure "Jaundice, Dropsy, Rheumatism, Fits of the Mother, Heart-burn, Agues, Fluxes of Blood, Internal Ulcers, Pains in the Head, Stone, Gravel, Palsy, Leprosy and King's Evil [Scrofula]", as well as "Burning or Malgnant Fevers". Tipping's Cordial Liquor was sold as a "wonderful and speedy Dissolvent of the Stone, which cures the Strangury and Ulcers in the Kidney and Bladder, and speedily relieves the Colick, Gout, Rheumatism, Palsy, and all Chronical Distempers" that also restored strength to "debilitated Limbs" and eased "all intermitting Fevers, Hectick and Consumptive Cases". Readers of the *Athenian Gazette* learned that "Mrs. Norridge, [who] now lodges at the Blew Ball in Exeter-street in the Strand" was selling an "infallible Powder for the Stone and Gravel" originally made by her father from a secret recipe that he bequeathed her.

There were no regulations about where these medicines could be sold, either. Anthony Daffy initially peddled his homemade Elixir Salutatis at Robert Clavell's printing shop in London. The "famous, little, ITALIAN BOLUS" that supposedly cured "Venereal Disease" was available at the "Flaming Sword the Corner of Russel Street, against Will's Coffee-house, Covent Garden". A "Chymical Liquor" for treating eye problems was available by appointment only, "at the Gentle woman's at the Two Blue Pots in Haydon-Yard in the Minories, London".

A stroll through late seventeenth-century London offered endless opportunities to buy a cornucopia of healing concoctions from printers, booksellers, bakers, hosiers, toy sellers, chocolate makers, postmasters, grocers, and linen drapers, at coffee houses and taverns. Enterprising nostrum makers also sold their wares in other cities. One advertisement assured readers that its elixir was "sold by some one Bookseller in most Cities and great Towns in England, and in the chief Coffee-houses in and about London". Customers could purchase Dr Bateman's cure for gout in Newcastle-upon-Tyne, Northampton, and London, while those in London, Bristol, and Gloucester could get "The Famous Specific INJECTION or LOTION" which "intirely [*sic*] destroys and carries off all Venereal Contagion".

There were fortunes to be made in this developing medication market, and plenty of people made them. In the eighteenth century, one of the most famous purveyors of quack medicines was Joshua Ward. (I am using the term "quack" to refer to medication that is not made by or for physicians and/or apothecaries. It encompasses medication that the seller truly believes will ease suffering or cure ailments all the way to compounds that the seller is using solely to get money.) Ward's personal story illustrates how volatile the social, economic, scientific, and medical systems were in the first half of the eighteenth century. Legend has it that as a young man, he became an MP under dubious circumstances. Ward duly presented himself to take office, the powers that be ruled that he had committed fraud, and Ward hotfooted it to France before he could be arrested. What is better documented is that Ward found a place among France's abundance of quacks, frauds, and hucksters. Two Englishmen, Sir Thomas Robinson (a future member of Parliament) and John Page (a future Secretary of the Treasury), met Ward and his medicine in Paris. Shortly thereafter, Ward was pardoned. He returned to England, bringing with him the first version of his miraculous elixir.

In London, Ward wasted no time advertising Ward's Pill and Drop. He even hired people to claim to have been cured by them. Between advertising and word of mouth, Ward very quickly built a very large, very enthusiastic client base. He opened a store for selling his goods (as opposed to a shop, where he would have made and sold them). Customers flocked to it and a cadre of women sat outside it crying up his goods. His clients included Lord Chesterfield of furniture fame, the author Henry Fielding, the historian Edward Gibbon, and George II. (The story is that His Majesty dislocated his thumb and Ward wrenched it back into the socket; after both pain and temper subsided, the king endorsed the charlatan.) Unsurprisingly, Joshua Ward very quickly became famous and rich. He expanded his products to include a "blue pill", a "purple pill", a "red drop", a "white drop", haemorrhoid cream, headache medicine, the lovely sounding "liquid sweat", a "Purging Powder" for swollen limbs, eye drops, and a medicine for treating gonorrhoea that was injected. Injection needles were not nearly as small or as sharp as they are today, and plungers did not depress as quickly; an injection would have felt not unlike being stabbed with a blunt pencil. That this medication had a market testifies to Ward's reputation (and people's desperation).

Ward cannily legitimized himself with ostentatious performances of benevolence. Promised George II's patronage for returning the royal thumb to the royal socket, Ward had himself installed in His Majesty's almonry office so he could dispense his cures to the poor for free. With this glow of charity about him, Ward then used his financial success to establish a hospital for the ailing indigent and a house within the City of

London, on Threadneedle Street, to serve more as a clinic. As "Philo-Chemicus" ironically explained, "pretending a Public Charity, or the universal Good of Mankind" conferred the "Authority" to practise medicine even in the absence of training or a medical licence. Or as Ward often pointed out, he could not be a conman because a conman would never be so generous with money (and product) to the poor.

There were plenty of sceptics, especially about the Pill and Drop. The poet Alexander Pope mocked Ward for having neither proper training nor reliable medications. "He serv'd a 'prenticeship who sets up shop," Pope wrote, but instead of serving an apprenticeship, Ward experimented on test subjects: "Ward tried on puppies and the poor, his drop." Sir Charles Hanbury Williams, a member of Parliament with a larger talent for invective than for oratory, wrote two bitter poems in which he satirized both man and Pill. Poetry was a popular medium for seventeenth-century and eighteenth-century critics, and poetic attacks appeared in the newspapers of the day. One poem in the *Universal Spectator* addressed "Egregious Ward", sneering "you boast with success sure, / That your one drop, can all distempers cure" before listing cures that it had failed to perform. In another issue, an untalented but enthusiastic critic of Ward's wrote, "'Tis plain Ward's nostrums arn'td [*sic*] dispensed for money all," warning readers that "I fear, instead of Tartar's Cream and Scammony, / You'll catch a Tartar [a shrew], and find all a Sham on ye." The *Universal Spectator*'s readers were neither the first nor the last to be unpersuaded by really bad poetry, but the effort was there to change Ward's medicaments from fashionable to embarrassing.

A more substantial attack was launched by physicians, apothecaries, and Fellows of the Royal Society. After all, the phenomenon of Joshua Ward and his miraculous Pill and

Drop was exactly the kind of phenomenon that the Scientific Revolution aimed to eliminate. Ward kept the formula for his Pill (and everything else he sold) secret, but the New Science valued knowledge sharing. Ward made extravagant claims for the Pill's efficacy, but medical professionals had established that there was (alas) no universal panacea. Ward made a virtue of his ignorance, proclaiming his complete lack of medical and chemical knowledge as loudly as he proclaimed the virtues of his medications. Secrecy, illogic, and a complete disregard for, if not active devaluing of, hard-won facts, insights, and methods, was bad enough. The real danger that Ward's Pill posed to those who took it was appalling and horrifying. Starting in 1734, physicians, apothecaries, and other proponents of the New Science applied the scientific method to Ward's most popular medication to establish what it was, how it worked, and whether it was safe.

Not coincidentally, the popular press was the primary field of combat. In July 1734, early in Ward's ascent to fame and wealth, a short piece in the *Gentleman's Magazine* warned readers that Ward's secret formula was not a secret to the physicians and furthermore, the reason why the public could not get such a treatment from them was because men like Sir Hans Sloane thought that it hurt rather than helped patients. When in November and December the *London Evening Post* and the *London Daily Post* offered testimonials of the Pill's efficacy, other newspapers printed furious rebuttals. The *Weekly Oracle* began publishing a series of long articles by an unnamed physician in which the author examined the credibility of Ward's claims. The "Operation of all these *Antimonial Vomits*, is so very precarious, that no Man can be absolutely certain how they will turn out," the anonymous author warned, underlining the "Hazard and Uncertainty of their Operation, and the manifest

Danger that must frequently attend upon many of those who think proper to venture on them". The author acknowledged that he had "given many Pounds of *Crude Antimony*", but he distinguished himself, a medical professional, from those who knew nothing of the "Remedies they recommend, much less of the Diseases for which they recommend them, and less still of the human Frame which is to grapple with them". In a subsequent instalment, he assessed which diseases the Pill really could cure, quite nicely informing his readers that Ward's Pill could cure none of them.

The unknown author's goal was to educate the public to protect them from the worst dangers of Ward's Pill. Having pointed out that it was dispensed by someone who knew nothing about chemistry, disease, or anatomy, he listed the types of people who should not under any circumstances take it. Or, as he put it, "I thought I might do an acceptable Piece of Service to the Publick, if I laid down some Cautions in Reference thereunto." He also made it difficult for Ward to object to these "Cautions" by claiming that the "Dispenser of them to necessitous Persons, will not be displeased" because of course, "the more Good he can do with them, without any unhappy Accident supervening, the great Applause to him, as well as Esteem for his Remedies". This attempt to limit the number of people injured or killed by Ward's Pill addresses the self-interest of both parties involved in the medication marketplace: the seller (Ward, in this case) and the buyer (the patient).

A more aggressive, protracted campaign was waged by a partnership between the *Grub Street Journal* and several members of the medical (or medicinal) community. A document in the British Library's Sloane Collection reveals that as early as 13 August 1734 a team, perhaps of physicians, apothecaries, or Fellows of the Royal Society, performed several chemical analyses of Ward's Pill, including examining the crystals under a microscope, heating a ground pill, and combining a ground pill with another substance to catalyse a reaction. "The conclusion is this," they reported, "that it can be nothing but the glass of Antimony powder'd, colour'd wth cinnab[ar]: and made up with Gum: for no praparation [*sic*] of Antimony but the glass will do to so small a dose." Just to make absolutely sure, however, "We have try'd ours uppon several people, and find our perfectly to coincide in the operation, & to produce the same effect." (Remember that Ward's Pill usually made people vomit or gave them diarrhoea.)

Later that autumn, Dr David Turner performed his own analysis and wrote a lengthy letter to the president of the Royal Society, Dr James Jurin. Like Philo-Chemicus and the anonymous writer in the *Weekly Oracle*, Turner proved that Ward's famous Pill was a form of antimony that was known to be toxic. Deeply concerned that the public was paying for the privilege to poison themselves, he begged Jurin to pressure the government into banning the use of antimony in over-the-counter medications. When the letter did not work, Turner turned to the press to alert the public and pressure Jurin. His treatise, *The Drop and Pill of Mr. Ward, Consider'd*, was subtitled *In an Epistle to Dr. James Jurin, Fellow of the College of Physicians, and of the Royal Society*. Dr Jurin and the Royal Society did not take any action, but the *Grub Street Journal* did. The newspaper printed excerpts of Turner's letter to Jurin, including twelve case histories of people who had taken violently ill or died from the Pill. Two weeks later, the newspaper featured a letter of support from another physician, and two weeks after that, a second one. Admittedly, this issue first printed an exuberant defence of Ward and the Pill (even eighteenth-century newspapers ran on advertising), but by following it with the serious, knowledgeable letter, the editor gave the professionals the last word. When Ward provided a rebuttal, it consisted of *ad hominem* attacks on Turner, assertions that he had evidence supporting his claims (but none of the evidence itself), excuses (some of those who died were already "at the point of death", for example), "what-about-isms" such as, "He says nothing of all those incurables who die in the operation of his own medicines, and those of his brethren, tho' ordered by the most learned and experienced," and most ridiculously, that he did not have time to write a "true and candid relation of the good and bad effects" of his medications. Mr Ward might indeed have been busy, but he was not too busy to write a riposte taking up one full sheet of the newspaper and two-thirds of another.

Enter Joseph Clutton, apothecary. He was born in Worcestershire in 1695 and apprenticed at the age of fourteen to the apothecary Benjamin Morris. Not unusually, he married the daughter of another apothecary, Richard Morris, who was likely a relative of Benjamin's. Mary Morris Clutton was not only well educated but also trained as an apothecary. Joseph stipulated in his will that she should take over the business until their son, Morris, was of age, and letters to her were addressed to "Mrs Clutton, apothecary". Joseph Clutton was interested in chemical medication from the start of his career. In 1730, he bought from two "druggists" their lease to a sizeable property in what would become Coldbath Fields, part of Clerkenwell. The lot included "a Shed a Countinghouse" and, most desirable, an "Elaboratory with other conveniences for the making of Chymical preparation", such as "Potts Glasses Furnaces Grates Barrs Goods Utensils & things". Clutton also purchased the right to build a street directly to his laboratory from the main thoroughfare. In 1739, he became the supplier of the chemical bases for medications to Hampshire County Hospital. (When he died in 1743, Mary took over the contract.) His apothecary off Dorrington Street in Holborn was at the sign of the Bell and Dragon.

Although in time Clutton's son, Morris, and Clutton's apprentice, Thomas Corbyn, would turn the business into a chemical medication supplier, Joseph Clutton himself was a sincere adherent of the New Science, passionate about the developing field of chemistry and its ability to benefit humanity through medication. His attack on Joshua Ward and Ward's Pill was intended to destroy Ward's credibility and promote the New Medicine and the New Science. He was pushed to act after Ward brought suit against the *Gentleman's Magazine* for defamation instead of producing evidence to refute their claims. That act told Clutton that Ward was more interested in silencing his critics than he was in vindicating himself with evidence. As Clutton wrote, initially Ward "seemed inclin'd to appeal to the Publick, if not upon medicinal Theory, yet at least upon Fact and Experiment: But he quickly grew weary of this manner of Trial". So the apothecary did what any modern, scientific thinker would do: he collected data to establish the Pill and Drop's safety and efficacy, and he conducted a series of experiments on Ward's Pill to determine its ingredients. He published his findings in 1736 as *A True and Candid Relation of the Good and Bad Effects of Joshua Ward's Pill and Drop.* He had a great deal to say. It was 122 pages long.

Ward was not just a competitor of Clutton's: he challenged in a very high-profile way the nascent medical establishment and even more alarmingly, the entire Scientific Revolution. Ward's claims and strategies in the late 1730s touched on all the points that proponents of the New Medicine were trying to overcome. Ward argued that there was nothing special about medical education or training, and made a virtue of his lack of either. He contended that there was nothing mysterious or special about effective medications. He used chemical medications like the professionals did, but discredited their analyses and therefore the scientific method. Ward ostentatiously cared for the poor, filling a gap created by the commercial system that physicians and apothecaries were building. He claimed his own data and brought out his own disinterested witnesses, some of them of high rank. In short, Ward was saying that chemical medication worked better than that crazy organic stuff that women made, but you did not need to be one of the elite in education or apprenticeship to make or use it. The knowledge was available to anyone. Furthermore, you did not need this testing mumbo jumbo to come up with great cures.

A True and Candid Relation was not Clutton's first time in the lists against Ward, although in their first engagement he signed himself Misoquackus (he might also have been some of the other anonymous critics of Ward and his medication). In A True and Candid Relation, Clutton states that since Ward has declined to defend himself and is instead suing the Grub Street Journal a second time ("yet he has given us no Notice of Trial" [emphasis added]), he, Clutton, "shall therefore proceed in my first innocent Intention of informing Mankind, by Examples and Facts, how violent and dangerous his Medicines are". Clutton's attack was very simple: Joshua Ward did not care about anyone's welfare, he only cared about making money; he aggressively preyed on people, especially the poor; and he sold poison under the name of medication. Clutton's method for making this argument was a model of the Scientific Revolution's thinking: he destroyed Ward's credibility while justifying his own, and he discredited Ward's evidence while providing an avalanche of his own, reliable evidence.

It was a duel, Clutton brandishing the New Science against Ward wielding emotion, and as duels are, it was personal. For one thing, Clutton understood that he had to be the Scientific Revolution's ideal, the Modest Witness. For another, Ward had gone after him by countermanding Clutton's instructions to a mutual patient and sending people to Clutton with stories of miraculous cures from Ward's treatments. So Clutton emphasized his own scientific character: disinterested, working for the good of humanity, transparent with method and conclusions, knowledgeable, and reliable. Had Ward's Pill "done, or would it do, but ever so little more Good than our common Medicines, I should have been as ready an Advocate in promoting it", he contends, but "I cannot, however, be just to my first Principles, of being *honestly concerned for the publick Good*, if I cease to publish some farther Discoveries on this Head" (original emphasis). He even provided the recipes for one of his treatments. As for Ward, Clutton showed him as greedy, predatory, ignorant, and secretive, taking care to use Ward's own words to do so. Over and over again, for instance, Clutton recounts how someone has become ill from one of Ward's concoctions but is urged by Ward to take more of it.

Character alone is not evidence, however. At best, it can establish or reinforce the credibility of the evidence. In addition to proving throughout *A True and Candid Relation* that Ward was trustworthy, credible, and interested only in the good of humanity (in other words, the ideal scientist), Clutton provided a tsunami of evidence. One kind was case studies: accounts of real people and what really happened to them when they took Ward's medications, especially the Pill. To demonstrate a scientific willingness to entertain data that challenged his conclusions, Clutton provided several accounts that people gave him of their own cures. "A poor woman came" to his shop, he reports, "to tell me she was better by taking of Ward's Medicines" and a man from Greenwich tells Clutton that Ward's "Medicines" had cleared up "scorbutick Breakings out about his Arms". Clutton then called that evidence into doubt. The "poor woman" admits that "some Woman persuaded her to come and tell me so" and the man from Greenwich "evaded that Question" when Clutton asks "who sent him to me". After Clutton says that "those who sent him were much in his Debt, for coming so far in such bad Weather", his visitor "compos'd his Countenance, and took Leave". In some cases, Clutton showed how signed affidavits attesting to Ward's miraculous cures were falsified, compelled, or bribed into existence.

Then there was the case of the woman who "had a Wolf in her Stomach, and was used to devour four or five Pound of raw Meat at a time, but was now cured of it", a story that a friend of Clutton's was told when the friend went to Ward's office. "A Surgeon of Note had it from *Ward* himself, that she eat two Legs of Mutton at a time," Clutton writes, before calculating the weight of the average leg of mutton in London. He concludes, tongue firmly in cheek, that "It must not be a small Wolf in her Stomach, which could devour either ten or five Pound of raw Flesh, and therefore this must be a very remarkable Cure."

In contrast to these dubious accounts, Clutton provided two large sets of cases whose sources were reliable, that is, verified either by Clutton himself or by other credible authorities. These cases were appalling, recounting profound suffering, permanent disability, or death. Clutton himself dealt with people who were determined, in his opinion, to kill themselves by using Ward's medicines. "Three or four of my own Acquaintance" took Ward's medicines "for slight Ailments; two of whom had like to have been kill'd, and apply'd to me for Assistance."

Friends and colleagues shared their experiences; even Clutton's laboratory assistant had been devastated by Ward's treatments.

Also helpful, the *Gentleman's Magazine* had been collecting and publicizing horror stories, which Clutton retold (giving full credit to his source). For example, Hester Staps (aged forty-five) took Ward's Panacea to treat her "small scorbutic pimples, which used to break out spring and fall". The first dose "vomited and purged her time beyond numbering", but since the Panacea was supposed to cure vomiting and diarrhoea, Hester took two more doses. She lost her appetite and developed a migraine, fever, respiratory illness, "continual griping pain in her bowel", and depression (no wonder), after which "a most violent leprosy" broke out "all over her body" and she began to shed "scales". Finally, she "miserably wasted away" and died. John and Daniel Wooten, both in their mid-30s, took Ward's Drop for nearly two months despite getting sicker, and Daniel then took one of Ward's Pills. Unfortunately, "Daniel's pill burst a vessel within him in the working, and forced up a quantity of blood." He and John survived only another five days. James Frettwell, a forty-nine-year-old carpenter subject to dizzy spells, took Ward's Pill and then his Liquid Snuff, developed swelling in his face, and became violently insane before dying several days later. "Mrs. Magee's daughter", a five-year-old with a rash, took three doses of Ward's medication and was dead by morning, and Mrs Riely's three-month-old died in her arms immediately after Ward gave the infant Liquid Snuff.

There is something sensational about this kind of evidence, however, and Clutton took care to it shore up with other, less titillating kinds. In response to Ward's taking credit for a drop in the mortality rate for 1734, Clutton provided an analysis of the Bills of Mortality that showed no change in the mortality rate to those who suffered from the ailments that Ward claimed to cure. Corroborating experts appear throughout *A True and Candid Relation*, such as Pierre Dionis, the renowned French anatomist; Nicolas Le Fèvre, brought to England by Charles II for his expertise in chemical medication; and Nicholas Lémery, an apothecary and chemist who was one of the first to propose the idea of acids and bases. Clutton also refers to other experts at the time that he was writing, such as the surgeon William Cheselden and the physician Richard Mead. Others, like J. Mason, surgeon, or Francis Dalby, apothecary, reveal the wide network of knowledgeable apothecaries and physicians within which Clutton worked.

Because the chemical composition of Ward's Pill was vital to the argument against it, Clutton also walked the reader through the chemistry and the procedures by which he analysed the Pill. He defines the kinds of antimony and explains, one by one, how each is made. He proves the presence of arsenic in Ward's pills by explaining that:

> Men of Learning generally agree, that of all the Ways to find out the Virtues of Medicines, the two following are the chief; the first is to strictly observe their Effects, the other, to trace them back to their component Principles.
>
> The former of these two Ways we have gone through in the preceding Cases, and laid before the Reader both the good and bad Effects of the *Pill* and *Drop*. By comparing these with what Authors of the greatest Credit say of the Effects of Arsenick, I think is the fairest Way to prove that Arsenick is in *Ward's Pills...*

This is not prose for specialists but for the lay reader, the people who are taking Ward's so-called medicines on the strength of word of mouth and perception rather than on evidence and reason. Consistent with the founding principles of the Scientific Revolution, *A True and Candid Relation* states what it is going to do and how it is going to do it. "I shall therefore observe this Method in the following Sheets," he announces early on, and then provides a numbered outline of his argument. There are guidelines along the way, such as "all things divided into three classes", "I shall now in the 3d place shew, that this *Nostrum* is a poison," and "Having informed the Reader what *Cobalt*, *Arsenick* and *Zaffer* are, we shall proceed." All vocabulary is defined, all processes are explained, no foreign language (ancient or modern) is used, and every step is identified along the way.

The full breadth of Clutton's argument appears best by looking holistically at *A True and Candid Relation*. Taken together, the sections establish that Ward does not have a consistent recipe for his medications. Early on, Clutton shows that the Pill's weight varies greatly, which means that production is not standardized. Part I of the treatise, Sections I and II, show how Ward uses antimony in his medications. The case studies have similar symptoms (primarily vomiting and diarrhoea); Clutton's chemical analysis reveals the presence of antimony. In Section III, the centre or fulcrum of the argument, Clutton uses Ward's defamation suit against the *Gentleman's Magazine* to justify the extensive reprinting in Parts I and II. He also uses the suit to protect the never-before-seen argument in Section I and Section IV, its parallel in Part II. Part II of the treatise, Sections IV and V, uses the same structure and method as Part I, except to show how Ward also uses arsenic in his medications. The case studies in Part II share a set of symptoms, but in addition to vomiting and diarrhoea, sufferers experience muscle spasms, swelling, numbness and paralysis, paranoia, and madness. If treated for open wounds or intestinal disorders, the sore or excretions acquire profoundly foul smells. (People who lived with pit toilets, chamber pots, and horse dung in the streets were driven out of the room by the odour. That's bad.)

What does this mean? For one thing, apothecaries and physicians knew that antimony and arsenic are both very, very bad for people, probably (hopefully) one reason why Paracelsan medicine never really caught on. The "five known preparations of *Antimony*", Clutton explains, "for their violence and harsh manner of working, are but very rarely prescribed by physicians". One grain of "Algarott", the fifth preparation, "is so rugged and harsh, working upwards and downwards, with so much pain, and such uncertainty with all, that it is seldom or never prescribed". Yet Ward used "Algarott" in his Purging Sugar Plumb, a medication designed for children. As for arsenic, Clutton provides only case studies of fatalities, including post-mortem results. It was alarming that Ward was using these metals at all. It was even more alarming that he did not control how much of either he used, and that there was very little predictability as to which he was using. In other words, there was no consistency in his medications. He did not have a recipe. If there was no recipe, then Ward was just throwing cheap materials together and selling them under assorted titles. Clutton says as much – in fact, he says that quacks do it all the time. Antimony is cheap and easy to procure, so "Ignorant and bold quacks generally make these articles the basis of their packets." He explains precisely how the profit margin works: metals like antimony "are all suitable for Quacks in that respect, the Chymist selling after the rate of 40 grains of the two latter a penny, and 480 grains of the 3 former for the same money. So there are 480 doses for a penny." Antimony is also useful because ingesting it makes people so sick that they "feel they have somewhat for their money".

Clutton's ally, the *Grub Street Journal*, used *A True and Candid Relation* to continue attacking Joshua Ward. An unknown poet wrote:

> In this bright age three wonder-workers rise
> Whose operations puzzle all the wise.
> To lame and blind, by use of manual flight
> Mapp gives the use of limbs, and Taylor sight.
> But greater Ward, not only lame and blind
> Relieves, but all diseases of mankind
> By one sole remedy removes, as sure
> As Death by arsenic all disease can cure.

"Mapp" refers to Mrs Sarah Mapp, who had a short but highly successful career setting broken and dislocated bones, and "Taylor" is Dr John Taylor, who achieved fame as an oculist. It was not quite fair to attack Mapp like this, as unlike Ward's and Taylor's businesses, bonesetting had no room for quackery: either the bone set or it did not, and everyone could tell just by looking at the injured person. Unsurprisingly, there was a great deal of reluctance in some quarters to trust her, not least because she was female, highly successful, and ostentatiously enjoying it. For example, Hans Sloane was very cautious about acknowledging her effectiveness even when presented with irrefutable evidence of it. In a letter dated 18 November 1736,* he corroborates an

---

* The spelling and punctuation of this letter are so idiosyncratic that I have modernized and corrected them enough to be comprehensible to a modern reader who is not Sloane's correspondent.

account that his correspondent had heard about Sloane's step-granddaughter:

> M^iss Isted was carried to Epsom, to Mrs Mapp for many disorders in her joints being awry. This [journey] was kept a secret from me lest I should be averse to such a trial, [Miss Isted] being my late wife's granddaughter. I saw for that reason none of the operations nor meddled save in taking care of her after she came up with Mrs Mapp on account of a dangerous colick which I do not attribute to ye operator. After she was free from the danger of the colick, I left her to Mrs Mapp's care. I believe she [Miss Isted] had received benefit by her, but she is not yet got out of her [Mrs Mapp's] hands. She herself told me she gathers strength every day & her friends that she is straight although one leg is shorter than the other by being 14 years contracted. Mrs Mapp by relation of credible persons has replac'd bones both in the ankles & shoulders wch have been out of their places & I think is now in London removed from Epsom. You have here a plain short account of my knowledge & belief in this affair who am
>
> Your most obedient & most
> humble servant
>
> Hans Sloane

He is fair: he speaks from medical expertise, not prejudice, about Miss Isted's colic, and after he has treated her for the colic, he allows Mapp to take over the case again. Still, he is careful not to endorse anything; in the original document, the word "certainly" is crossed out and replaced with "by relation of credible persons". Disrepute clung to Sarah Mapp, and she remained the sort of practitioner who was satirized in public and consulted by respectable people in secrecy, if at all. Sloane's letter preserved the distinction between himself, a real physician, and her, a bonesetter. After all, many women learned of necessity to set bones – it was the kind of thing that happened in families. Sarah Mapp had the nerve to attend strangers and charge for it.

The line between quacks and legitimate practitioners was easy to blur. William Hogarth, recognized in his time as one of the most important visual artists of the period, skewered physicians as well as Ward, Mapp, and Taylor in his print "The Company of Undertakers" (see plate section, p.6). Taylor has one eye closed and is holding a cane with a Freemason's eye, as if his eye has jumped from face to cane; Mapp is bloated, cross-eyed, clutching a bone for a cane, and dressed like Harlequin; and Ward has half his face shaded, also like a carnival figure. Both men are snuggling up to Mapp, who is staring directly out of the frame at the viewer. Hardly complimentary. However, below them, unobserved by the trio, is a cluster of twelve wigged, cane-wielding physicians. They are facing in all directions as if in confusion; most of them are fleshy, except for two who look rather cadaverous. Hogarth has framed the whole scene with a coat of arms and unscrolled across the bottom the motto *Et Plurima Mortis Imago*: "The many images of death." For Hogarth and popular conception, everyone in the frame, quack or professional, grotesque or respectable, is an undertaker, an image of death – someone who helps the ailing and injured into their grave.

Hogarth was not wrong that, on the whole, neither the professionals nor the quacks offered anything truly curative.

Nor was he wrong that commercial medicament dealers were taking advantage of a market that the physicians and apothecaries were busily establishing for their own gain. Clutton's *A True and Candid Relation* cut this Hogarthian knot. He distinguished between the fraud and the professional by showing the audience exactly how scientific medicine worked so they could compare it favourably to the often illogical and never proved promises of individuals like Ward. He also showed the difference between treatment by people you know and the invisible, unavailable quack. Reliable, effective medication was medication obtained by someone the patient could meet in places associated with that someone's authority. In the case of physicians, prescriptions for medications were obtained through an in-person visit, so the sufferer could trust that they had been seen, literally and medically, and they could trust who saw them, the physician. Patients who so desired and could afford it could be attended at their own home. One frantic writer warned his physician that:

> If this note finds you at home, I will call on you now. If it does not, I must beg you to us, to morrow morning, the person on whose account I have already given you so much trouble: and who is so very dear to me. Her pains are returned upon her, and have been for these 3 days, with cruel violence.

House calls were private only in a relative sense. The patient was attended by family, servants, sometimes friends, and sometimes other medical professionals. Visits could be followed by still further visits, but as previous chapters also demonstrate, a great deal of medical business was conducted by correspondence between the patient's representative and the physician or apothecary. One son wrote about his mother to the attending physician, and the message was delivered by the patient's daughter. "Sir", the anxious letter began:

> My mother is much the Same as when yr honour See her. She had a very Bad night, & is Exceeding sick att her Stomach, She swells much in her Bode, & take but Little water, & her Breath, for most part is very Short. She take the oyle's Surrup often, but has not took of the other Since yesterday in the After noon, it is so bitter, & heats her mouth, which is always dry. i [*sic*] beg Sir you will tell my Sister what you think of my poor mother and yr honour goodness to her very much adds to the many obligations already laid on Your humble Servant Jas Keill

Mrs Keill's health was a matter for the household, involving two children who knew quite a lot about her. They were responsible for conveying and receiving all medical information.

Professionals also acted in the public eye. They made use of their journeys from patient to patient to advertise their skill and authority. Clothes, servants, servants' clothes, and equipage were all displays of success. Dr John Radcliffe, the most successful and respected physician of his day, carried a gold-headed cane. When he died, the cane went to Richard Mead, his protégé, who by then was the most successful and respected physician of *his* day. The cane was handed down from one generation to the next, until it was finally retired as an object sacred to the RCP. Physicians also made use of a public forum to display their skill and knowledge. By the early eighteenth century, medical consultations with physicians often took place in coffee houses. Grouchy Dr Radcliffe frequented the Bull's Head Tavern; his successor Dr Mead preferred Batson's Coffee

House. Part of what made Sarah Mapp so outrageous was that she held consultations at the Grecian Coffee House.

Occasionally, depending on the physician and the apothecary and the nature of their working relationship, an apothecary might spend some time with a physician at the coffee house. As Clutton's case studies indicate, by the 1730s apothecaries could also be sent for, like physicians. Primarily, however, apothecaries were tied down because they had stock, the material for making medications. So apothecaries used their shops to proclaim their honesty and skill, and their medicines' efficacy and safety. The development of the shop as a place of commerce, sociability, and authority was central to the development of Western capitalism. The apothecaries were part of this trend as medicines and the substances to make them became commodity and capital.

Apothecary shops generally shared certain features. The shop was divided into two zones: the public, commercial zone where people handed over prescriptions and talked with the apothecary and assistants, and the private, operating zone where most of the ingredients, including the really rare ones, were stored and where all medication was made. (This operating zone was one of the reasons why early chemists met in apothecaries' shops.) From the middle of the seventeenth century, apothecaries developed the public area into a place that testified to their skill, knowledge, and honesty. There were boxes, jars, canisters, and bottles of all sizes, designed for a wide range of functions and made from an array of materials, from ceramic to wood to glass. These were displayed on floor-to-ceiling shelves all around the shop and sometimes even over the door. The front room of one late seventeenth-century apothecary held twelve barrels, one large box, more than 160 small boxes, thirteen syrup glasses, 130 gallipots, another eight regular pots, sixty-five bottles, and eighty-two parcels of medicinal substances. Some apothecaries had equipment in the front room for performing simple procedures, such as making pills or grinding up ingredients, giving customers a sense that the apothecary had nothing to hide and impressing them with his skill and knowledge. As historian Patrick Wallis explains, the "design" of the shop "functioned alongside other contemporary sources of customer reassurance, such as the vendor's reputation". In contrast, the majority of quack medicinal cures were not sold in person by their maker or in a place dedicated to healing the sick.

It might seem paradoxical, but the unseen space or spaces, such as back rooms or yards for brewing the really noxious stuff, were equally important for distinguishing professionals' endeavours and medicines. For one thing, it helped them differentiate their work from women's work in the kitchen, where everything was visible and known. For another, even though customers were forbidden these private, professional spaces (as opposed to the public, professional front room), customers knew where they were and who worked in them (see plate section, p.8). Advertised medications, on the other hand, were concocted often by unknown operators or in unknown places. The historian of science Penelope Corfield explains, "By the Georgian era, secrecy was a technique or hallmark of quackery." It was certainly both defence and offence in the drug market. Professionals who had their doubts about an advertised medicine could demand that the inventor of the medicine reveal their ingredients so others could test its true effectiveness. The ostensible logic was that if the medicine were legitimate, its purveyor would have nothing to hide; they would have no qualms about someone objective testing it to make sure it worked. Having the secret out, however, would significantly if not completely undermine the seller's control and profits.

One attack played out in the advertisement section of the *Daily Journal*. Apothecary John Moore's long advertisement for his highly effective "antiscorbutic" was followed directly by a highly sceptical, anonymous letter. "Whereas you seem to promise chymerical Wonders from your never-failing Antiscorbutic," the author sneered, "I beg you would satisfy the Public how the Scorbutic Salts are to be discharged without Sweats, Stool, or Urine, lest we should be apt to imagine this but an amusement to the world." He concluded, "Your Answer to this is expected, or your Silence will be taken as a secret Indication that you are conscious of your Error." It was the proverbial rock and hard place. Either John Moore explained all, allowing everyone else to make his medicine, or he said nothing, proving himself a quack.

Women were in an even more precarious position. Some widows were permitted to continue running their husband's apothecary shop, others were not. There also were women among the ranks of healthcare providers in London and in provincial towns. Readers have already met Mary Trye, Jane Barker, and Mary Morris Clutton, for example. The Cluttons' apprentice, Thomas Corbyn, finished his training with "Ann Burrell, Apothecary" and Mary Pardoe. Anglican bishops and archbishops also could license men and women, married and widowed, to practise midwifery, surgery, and medicine. Mr and Mrs Smith of Knutsford in Cheshire had three licences among them: surgeon and physician (him) and midwife (her). According to David Harley, in some rural dioceses women could only be licensed to practise midwifery, but in others women could also be licensed as surgeons and physicians. Episcopal licences were very difficult to come by; however, the Bishop of London licensed three women in 1727, one in 1728, and one in 1730.

Readers may remember from Chapter 5 that professionals, such as those belonging to the RCP, were generally content to let women care for the poor, the indigent, or the exceedingly rural ill. Women who posed more of a threat by treating paying patients, however, were not tolerated. Someone in Cuerden, Lancashire (population 401 in 1666) had a local woman, Mrs Garlick, charged with practising medicine without a licence. After twenty years of treating people in her London neighbourhood, Mistress Wilkenson was reported by a Fellow of the College. She was warned to stop practising and her husband had to pay a bond as surety that she would indeed cease. Sometimes, a case proceeded without the involvement of a spouse. Margaret Gibson, wife of Richard Gibson, was hauled up in front of the RCP's Board of Censors for giving medicine to Philip Chickering. She contritely confessed and was "admonished". Susan Lyon, the widow of an apothecary, was simply forbidden to practise and had to close the shop. Some women resisted. Accused of giving medication to ailing people, Elenor Woodhouse avoided punishment by disappearing into the depths of London, where she could not be traced.

Anything to do with medication was a particularly egregious transgression. Susannah Dry was imprisoned for selling it. The RCP warned Mistress Goodcole three times, then brought charges against her for making and distributing it. When the surgeon William Foster was charged with advertising his medication, he immediately pointed the finger at his wife, although he also contended that she had done a lot of good with it and "injured none". Like Mrs Gibson, the Fosters were "admonished" and ordered never to administer medication again. They also were forced to hand over Mrs Foster's recipe. A medication that safely and reliably cured people of chlorosis, a form of anaemia that turns people greenish yellow, was of great interest to the RCP.

Joanna Stephens was not nearly so tractable when it came to her safe, reliable cure for the stone. Ultimately, it took an Act of Parliament to get her to turn it over. In the mid-1730s, her medicine for the stone was sweeping London and Paris. A cure that did not have a mortality rate for an agonizing condition? Yes, please. Physicians, apothecaries, and medical-minded men on both sides of the Channel recognized that "Mrs Stephens Medicine for the Stone" could have a huge therapeutic impact. They also recognized that it posed serious financial and medical competition. Complicating the problem, Joanna Stephens's gender and seemingly unmarried state made her a threat to the gender definitions that they had been working so hard to establish. She was an outrage. She was alarming. She was also the perfect target: a woman – better, an unmarried woman – with no important social connections.

In 1737, Stephens acquired a nemesis: David Hartley. Hartley himself had suffered from the stone, so he knew first-hand how excruciating it was. He also had been cured of it – or so he thought – by Stephens's medication, so he also knew first-hand how effective it was. He wanted her recipe for the professionals, and he wanted Joanna Stephens out of the way. Chemical analysis might enable him to steal the recipe, but it would not put her out of business. His plan? Take the medication away from her by making it public property. In his first major publication, *Ten Cases*, he wrote bluntly that "My design in printing these Cases and Experiments is to engage the Publick to purchase the Discovery of these Medicines of Mrs. Stephens." In March, an announcement appeared in the *London Evening Post*. Signed by David Hartley, it proposed raising £5,000 from the public to give to Joanna Stephens in exchange for the recipe to her "Medicines for the Stone". He put it in terms of benevolence. In *Ten Cases*, he had taken a leaf from Joseph Clutton's book, writing: "But the Benefit which many Persons have plainly received from them in painful and dangerous Cases and my Opinion of their Efficacy in dissolving Stones of the Kidney and Bladder would render me inexcusable, if I did not use my best Endeavours to make them of general Service." In the *London Evening Post*, he explained that "at present the good Effects of 'em are confined to a few Persons only" because Stephens could not mass-produce it. "But there are in all Parts of the World great Numbers of Persons afflicted with the Stone, and whose Condition entitles them to all possible Compassion and Relief," Hartley lamented, and those poor souls deserved access to the medicine. Compassionately, Hartley acknowledged that the owner of the recipe should not "forfeit her whole Subsistence, by imparting a Secret of so much Importance to Mankind". This medicine, after all, was how Mrs Stephens earned money to live, but he promised in his proposal that Joanna Stephens would not be the loser.

It was a good performance, but in reality Stephens was far down the list of people or things that Hartley cared about. His proposal begins, "The Credit of Mrs. *Stephens*'s Medicines for the Stone, has of late been so much established, that many Persons of the first Distinction have thought them worthy of their Protection and Encouragement, and have desired that some Method might be take to make 'em public." Observe how Mrs Stephens does not come into this scenario. The passive voice of "has been so much established" means that someone else, not Stephens, has proved the medication's value. It is a disingenuous choice of tense as Hartley had recently published a tract with case studies – he means that *he* has "so much established" the medication's "Credit". Hartley goes from that point to what "many Persons of the first Distinction" want done with the medicines; Stephens's desires do not enter into his thinking. When Hartley does acknowledge her, it is to contend that she is incapable of handling the responsibility or demand ("herself not being able to make up for a large Number"). His connection to the medication, his perspective, others' interest, and Stephens's unworthiness all come together when he explains his purpose. "I have used my best Endeavours," he writes, "ever since I have been acquainted with the Value of the Medicines, first to dispose Mrs. Stephens to part with the Secret, and then the Publick to make her a proper Gratification for it." Stephens is the selfish loner; he is the benevolent, scientific member of society: "I have desired the Advice and Assistance of my Friends in every Step that I have taken."

As for the proposed "proper Gratification", it was designed to render her dependent. As Hartley explains a few months later in *An Account of the Contributions for Making Mrs. Stephens's Medicines Public*, Stephens would not actually receive the £5,000. That money would be given to one Mr Drummond, a banker. With every thousand pounds raised, Mr Drummond would buy annuities in the South Sea Company, and the portfolio would be held in trust for Stephens by a group of men, including Hartley himself and his partner, Peter Shaw. Once the trustees had raised and invested the full sum, Stephens would hand over her recipe for them to test, receiving in exchange the interest on the account until the professionals pronounced it cure or quackery. If they decided that the medicine was truly effective, Stephens would receive the annuities. If they decided that it was not, the money would be returned to the donors. What Hartley did not mention, of course, was that if the medicine were discredited and the money returned to the contributors, Joanna Stephens would be left with nothing except a bad reputation.

Hartley was trying to separate Stephens from her medicine, her income, and her good reputation, and to reattach all three to men. As he contended in the initial proposal, her medicine "may have many other Uses and Applications", but only "when in the Hands of Physicians" as a "leading Principle in their Reasonings upon the human Body". Hartley's approach also made the legitimacy of Stephens's medicine contingent on its meeting the standards of the New Medicine and the New Science. Her medicine had to be validated by observation and repeated testing by respectable, qualified, male professionals. Its operations had to be explained in chemical and biological terms. Hartley's plan was not about distinguishing male medicines from female medicines. It was designed to do the opposite: claim a successful female medicine for the male professionals.

Stephens held out against Hartley's pressure for months. After his initial salvo in March 1738, he kept her intransigence before the public eye with a stream of articles in newspapers. He appealed to the public to contribute to the fund to recompense her. He published the list of illustrious subscribers, implying that Mrs Stephens had some nerve refusing people of such importance. Sometimes he accused her of asking for an unreasonable sum of money, sometimes he said the amount was his idea and only fair. He rebutted the objections to his proposal in *An Account of the Contribution for Making Mrs. Stephens's Medicines Public; with Some Reasons for It, and Answers to the Most Remarkable Objections Made Against It.* He petitioned Parliament to pay the money, arguing that her medication was vital to the national good. When Parliament agreed to foot the enormous bill, he made much of that.

Stephens finally caved in December 1738. A collection of physicians, surgeons, and apothecaries recruited four test subjects who were suffering from the stone to take her medicine for different periods of time. To confirm the success of the treatment, groups of physicians examined each man, sometimes more than once. "Henry Norris, of Leather-Lane, aged 55", for example, was searched by "several Physicians and Surgeons" at St George's Hospital, who agreed that he had a bladder stone. After taking Stephens's medicine for four months, he returned symptom-free to St George's Hospital, where eight surgeons and physicians agreed that he had been rendered stoneless, so to speak. Sixty-seven-year-old Peter Appleton initially was searched by "Mr. Sharp" as well as "Dr. Pellet, Dr. Nesbit, Dr. Whitaker, and Dr. Hartley", all of them agreeing that he had a bladder stone. When he returned after five months, he had the pleasure of having a staff inserted up his penis and into his bladder by Mr Sharp and "Thirteen Physicians and Surgeons [...] at Child's Coffee-house in St. Paul's Church-yard", but "no Stone could be found". All four of the men passed their stone and achieved considerable relief. The cured patients then testified before a group of physicians and peers, after which the group approved Stephens's payment. Then the recipe appeared in newspapers from London to Edinburgh, slowly making its way into recipe books.

One of the test subjects was the same Mr Gardiner who appeared earlier in this chapter with Edward Nourse. In 1741, Nourse published a research paper on Mr Gardiner in the pre-eminent scientific and medical journal of the day, *Philosophical Transactions*. (His cover letter began, "Permit me to lay before you the bladder of Mr. Gardiner.") The article was also one of the very few of that time with an illustration: a fold-out engraving by the artist Elizabeth Blackwell, who had just completed the stunningly beautiful, magnificently erudite, two-volume book of medicinal plants entitled *A Curious Herbal* discussed in an earlier chapter. Nourse's medical and academic treatment of the case of Mr Gardiner and his bladder stone exemplifies the thoroughly scientific methodology with which David Hartley and his colleagues approached the testing and gathering of data. It also highlights what Joseph Clutton knew: that case studies made better drama than they did data. The *London Gazette*'s account of Mr Gardiner had a clinical tone, but it was a narrative all the same. All four case studies were narratives, with the gripping quality of a well-told story, however short.

Like the concurrent newspaper campaigns against Ward and his Pill, Hartley's newspaper campaign focused on the human angle. Unlike those efforts, chemical analyses appeared in publications that attracted a smaller audience. From mid-1739 to late 1740, four men – David Hartley and Stephen Hales of England, S. J. Morand and C. J. Geoffroy of France – not only deconstructed Stephens's medication but also established what made it effective and how those ingredients could be used more effectively. Morand and Geoffroy gave papers to the Académie des Sciences and then published their results. Hartley and Hales followed Clutton's example, publishing their results in tracts that also included case studies. Like Clutton, all four men could determine what went into the medication and deduce how it worked. Like Clutton, they all did so without the recipe. So much for Hartley's claim in 1737 that he needed the recipe to know what he was ingesting: "It is easy to see how much concerned I am to know what these Medicines are which I take daily." More evidence, if more were needed, that his relentless effort to pry the recipe for her medication from Joanna Stephens had to do with money and competition.

On the rare occasions that it is told, the story of Joanna

Stephens usually ends here, with the publication of her recipe and the destruction of her business. It is a convenient conclusion for anyone trying to lessen her significance rather than write accurate history. For one thing, although Joanna Stephens relinquished her recipe, it was not on Hartley's terms. Instead of waiting for the money to be raised by contributions, which was a dodgy proposition, Stephens was paid in one lump sum by Parliament after the results of the testing appeared in print. Furthermore, she was paid in pounds, not in stock. For another thing, Joanna Stephens did not vanish after this episode. Her name remained attached to the recipe even after it became public. It appeared under her name in recipe books, including apothecary books. Even Hartley continued to credit her for the recipe, recommending to "The Right Honourable The Lady Frances Shirley" in 1751 that she have "Mr Roberts an Apothecary in Pall Mall, at the Golden Mortar" make it for her. Furthermore, Joanna Stephens continued to practise medicine for another thirty-four years. The record of her burial on 17 November 1774 at St Paul's Church in Hammersmith identifies her as a "Doctress". It does not identify her as a widow. As the entries on either side of hers are for women identified as "widow", one of whom was also a "Shopkeeper", it seems that Stephens supported herself as a medical provider until her death at age eighty-two. What she did not do, however, was rebuild her business as a purveyor of over-the-counter medications – not under her own name, at least. If she sold medications, she must have used a different name or no name at all.

What, then, of Joshua Ward and his Pill? Financially, Ward cannily diversified his sources of income, building up his saltpetre and chemical works. Reputationally, he was unaffected by the efforts of men like Joseph Clutton. The satire *Siris Among the Shades* (1744) that teased Samuel Garth also took a moment to mock the "celebrated Pill and Drop", but that brief allusion was all. When Parliament finally passed a bill in 1748 to regulate over-the-counter medications, it excepted Ward *by name*. His fate was the opposite of Joanna Stephens's. Parliament named each of them in legislation, but to protect Ward and his business and to destroy Stephens and hers.

How to explain it? Ward was famous; Stephens was famous. His medication had an international following; so did her medication. The most powerful people in society, religion, and government believed in his Pill. The most powerful people in society, religion, and government believed in her medicine. There was ample evidence that his drugs did irreparable harm and could be fatal, but people kept taking them. There also was ample evidence that Ward knew the damage that his pills did, yet people continued to trust him. There was ample evidence that Stephens's treatment restored people to health, and that she was doing a great deal of good with it, yet she was stripped of her recipe and seemed to disappear. Chemists like Clutton established and published the recipe for Ward's Pills, but everyone kept buying them. French and English chemists found the composition of Stephens's medication and her recipe was made equally public, but it did not have the same staying power.

The explanation, of course, is that Joanna Stephens was a woman. Hartley and the British scientific community succeeded by taking advantage of that fact: they treated her recipe like domestic medication. They ignored or overlooked the fact that their analyses proved that Stephens's medication was more than effective; it was an intelligent compound, chemically speaking, with ingredients such as soap and eggshell that acted on the body in different ways to make the dose work. They certainly did not credit her with knowing chemistry or practising medical chemistry. Hartley and the men who followed him also used the ethos of commonality that underpinned recipe books to de-commodify Stephens's medication. They printed her recipe in the newspaper to show that unlike chemical medication (of course), Stephens's treatment could be made by anyone with a kitchen. At the same time, readers had to pay for the privilege of obtaining Stephens's drops, pills, or liquid from an educated, trained apothecary or physician. It was 1740, and medication from a professional was a commodity.

Actual regulation had a long way to go, but by the end of the 1730s, a nascent system was in place. Experts in chemistry and medicine ran human trials and chemically analysed a medication's components to establish its safety and effectiveness. They used print to distribute their findings, explain their views of the drug, and affect the market. From its inception, regulation was influenced by gender considerations. It also had a strong economic component. Approving substances expanded revenue opportunities and disapproving substances eliminated competition. The process strengthened the connection between medication makers (i.e. apothecaries), and medication prescribers (i.e. physicians).

If our ailing time traveller from Chapter 1 jumped forward one more century, from 1650 to 1750, they would hardly recognize the world. The concept of medication, the system for making it, the system for distributing it, and its components had changed profoundly. A time traveller jumping backward from the twenty-first century to 1740 would be appalled at what was offered, but the idea of paying for it? Well, of course.

Part Two

Ripples and Reflections

There is far, far too much to say about the global medication economy to attempt to cover it all, so Part Two has modest goals: to demonstrate how the developments and decisions of the seventeenth and eighteenth centuries (that is, Part One) manifest in the twentieth and twenty-first centuries, and to suggest ways in which domestic medicine has much to offer.

To begin this discussion, consider these three couples: insulin and lenalidomide (trade name Revlimid), thalidomide and aducanumab (trade name Aduhelm), fezolinetant (trade name Veozah) and sildenafil (trade name Viagra).

Insulin and lenalidomide are lifesaving medications, one for diabetes and the other for cancer. A century ago, insulin's developers attempted to ensure that only non-profit entities would benefit financially from insulin. One of the scientists, Frederick Banting, reportedly said, "Insulin does not belong to me, but to the world." His view has been shared by other significant contributors to medicine. Jonas Salk, who developed the dead-virus polio vaccine, and Albert Sabin, who developed the live-virus polio vaccine, refused even to attempt a patent.

Salk asked, "Can you patent the sun?" and Sabin announced, "It is my gift to the world's children." In a system where knowledge can be private property, however, recipes for medication (including chemical formulae and molecular structures) can be owned, guarded, bought, and sold regardless of what the medication does, including if it saves lives. And whoever owns the recipe controls the price and with it, access, which in turn affects health, suffering, and survival for many.

Despite Banting's highly quotable line, readers of Chapter 5 know that insulin became a golden goose for several pharmaceutical corporations, who had to be forced to make it available to more people. Government and popular pressure surely had a role in this change, but so did plain old competition – Walmart and Sanofi each announced their own brand of low-cost insulin in 2021. Eli Lilly's agreement in 2023 to lower prices included an immediate, significant price decrease for users of their insulin. In a use of punctuation deserving of a gold medal, reporters Jen Christensen and Betsy Klein neatly used a dash to show how Eli Lilly's benevolence covered and facilitated an attempt to acquire a larger share of the market: "The company said that its price changes should make a difference, but more is needed to help all Americans with diabetes – 7 out of 10 don't use the company's insulin." Joshua Ward would have been proud. Such a lovely cloak of benevolence with which to hide self-interest.

Like insulin, lenalidomide was synthesized to treat people with a specific disease, in this case cancer, and it is extremely effective. Also like insulin, in the United States it cost a fortune, had a history of frequent, large price hikes, and was unaffordable for many patients. The Celgene Corporation, the company that made it, charged US$215 per pill when it was approved in 2005; by 2019, the price was US$713. Bristol Myers Squibb bought Celgene later in 2019 and began charging US$763 per pill. The next year, the federal government began an investigation, which revealed that lenalidomide's price had been repeatedly raised solely to increase corporate profits, which in turn increased executives' bonuses. Summarizing the congressional testimony of Celgene's former chief executive officer, Mark Alles, Representative Katie Porter said, "So to recap here, the drug didn't get any better, the cancer patients didn't get any better—you just got better at making money." Well, of course. In the United States, pharmaceuticals circulate in a system where medication is first, foremost, and always a commodity, where its purpose is to generate income rather than to restore or preserve health. Like the scorpion in the fable, it is the nature of the system to pursue monetary gain regardless of the damage wreaked by that pursuit.

Unlike other countries, the United States has protected the market and the corporations often and paradoxically at its own figurative and literal expense. The Congressional committee found that "Celgene targeted the U.S. market for price increases while maintaining or cutting prices for the rest of the world. One presentation described the U.S. as a 'highly favorable environment with free-market pricing.'" In fact, the US market alone provided over half of Celgene's global profits between 2005 and 2019. Americans pay more for most medications than people do in other countries; in 2022, 10 per cent of Americans purchased their medications abroad in an effort to reduce costs.*

One reason for this difference is that other governments

* The Trump administration even proposed a programme to encourage Americans to buy medications from other countries rather than engage with pharmaceutical companies about the American market.

negotiate the price of medications with the pharmaceutical companies. Even after the US federal government authorized the largest purchaser of pharmaceuticals – itself – to negotiate drug prices with corporations in 2022, other methods for profiting from medications remained untouched. It took a year for those negotiations to begin and when they did, they involved only ten medications. Once the new prices were agreed upon, they were not scheduled to go into effect for another three years. Furthermore, the legislation limited negotiations to a certain number of medications each year, leaving the cost of hundreds of drugs unaffected. In short, the market that I described in Chapter 4 – that closed circuit between apothecary/drug maker and physician/drug prescriber, with minimal opportunity for interference or intervention by other parties – remained intact in the wake of the legislation. So did the decision to prioritize profit over others' quality or quantity of life.

Pharmaceutical corporations use other strategies to maximize their profit that descend from or are the same as those used in the seventeenth and eighteenth centuries. In the United States, every new invention, even if it is still at the idea stage, can receive a patent, a government assignation of ownership (see Chapter 2 on the privatization of knowledge). It is a crime to copy something under patent, such as a battery or a medication, but not after twenty years, when the patent expires. Joanna Stephens and quacks such as Joshua Ward knew that keeping secret the medication's formula is a key to financial success. If nobody knows how to make it, nobody can sell a rival drug. In 1725, an anonymous satirist explained that British medication makers:

purchase a Patent under the Great Seal for the Sole making, using or vending such a Machine, Preparation or Commedity [*sic*] [...] for a certain terms of Years, on condition that he then publishes it to the World; a sure way of saving their own Money, whatever the Subjects may suffer in the mean time for want of such a Publication.

In this millennium, pharmaceutical corporations follow this practice by tweaking the molecular structure of a medication to obtain a new patent on an essentially unchanged drug, thereby extending their monopoly and increasing profits. As Matthew Harper explained in *Forbes*, "drug companies file patent upon patent to try to extend the life of a single drug—turning to litigation to try to stifle generics". His language shows how supporters of Big Pharma view medication. About Bristol Myers Squibb, the company that bought Celgene and raised the price of lenalidomide to US$763 per pill, he wrote sympathetically:

> The company used court cases to delay generic versions of cancer drug Taxol and BuSpar. Generic Taxol was delayed for years. Keeping the copycats off the market made Bristol hundreds of millions of dollars and hurt generic drugmakers like Ivax and Watson Pharmaceuticals. But while this was going on, Bristol spent [US]$2 billion on a cancer drug from ImClone that wound up being delayed, watched the prospects for hypertension drug Vanlev sink on new clinical data, stuffed its inventory channels and saw generics eat away at its market share for many top drugs.
> The result: Despite its vigorous defense of its patents, Bristol Myers [Squibb] had to cut its earnings estimates for 2002 in half.

Oh no, not that. Harper does not address patient safety or benefit in his analysis, which makes sense if medication is a commodity used to make money, not something to benefit humanity. William Salmon had a point in 1698 (and Chapter 5) when he wrote that "a Monopoly [is] wonderfully prejudicial to the Lives, Liberties, Estates and Properties of the good People".

So let us return to the example of lenalidomide. After the congressional hearings and the committee's scorching report in 2022, the US price of lenalidomide increased. Why not? Even if any or all of the testifying executives felt shame or embarrassment when members of Congress showed them up as extortionists, there was no external incentive to act on those emotions and plenty of external incentive to keep on doing what they were doing. As a publication from the National Academies of Science, Engineering, and Medicine unironically put it, an "inherent conflict exists between the desire of patients (and society) for affordable drugs and the expectations of – as well as legal obligations to – corporate shareholders and other investors in biopharmaceutical companies". "Inherent conflict" indeed. Near the end of that congressional hearing, Representative Ayanna Pressley charged that "the lack of access to affordable lifesaving medicine is an injustice. It represents an act of economic violence, and an attack on the basic principle that health care is a fundamental human right." But healthcare cannot be a human right in a society that commodifies medication. A commodity *by definition* cannot be a privilege inhering in the human being or conferred by a government on all its citizens. Turning medication into something to be bought and sold in the seventeenth and eighteenth centuries involved a tremendous shift in values, and compromised medication's identity as a fundamental right of all human beings. When Dr Samuel Garth from Chapter 5 and those like him did not create access for those who could not pay, they reinforced that change.

Pharmaceutical corporations have a point that it takes a long time and a lot of money to develop a new drug. In the United States, approval takes one route, through the US Food and Drug Administration (FDA). In the European Union, there are four routes to approval, each for a different type of medication, circumstance, or combination of the two. Most drug approval processes involve four primary steps: the development of a drug, animal and clinical (human) trials by the drug developer to establish safety and effectiveness, expert review of the data gathered by the drug developer, and a decision by the regulatory agency. The process is easily summarized, less easily accomplished. The US National Institute of Child Health and Human Development states baldly that "The clinical trial phase can take years to complete." From some perspectives, the most significant thing about the drug approval process is that it is so complicated, time-consuming, and expensive. From another perspective, a long, rigorous process is warranted because history is replete with pharmacological disasters that damaged, destroyed, or outright killed adults, children, and foetuses.

Thalidomide is a global example of that peril. At present, it is used to treat leprosy and cancer, but its first appearance on the medication market in the 1950s was catastrophic (cancer researcher A. Keith Stewart called it "Shakespearean in scope"). Thalidomide was developed in the early 1950s by Chemie Grünenthal, a West German pharmaceutical company, as a sedative, and it was prescribed primarily to women. (This phenomenon is another instance of medication used to silence women, but that issue is much too large to be handled properly in this book.) According to Chemie Grünenthal's extremely inadequate testing, thalidomide had no side effects and could not be overdosed. The accidental discovery of additional uses, especially for morning sickness, made the drug even more lucrative. Companies around the world quickly bought the rights to sell the medication in their own markets. For the purveyors of thalidomide, it practically rained money. At the same time, two troubling phenomena appeared. The first was that adults who took thalidomide for a long time developed nerve problems in their extremities. The second was that physicians observed a shocking rise in the number of babies born with underdeveloped or missing arms, legs, feet, toes, hands, and fingers.* According to Sir Harold Evans, "Depending on the day it was ingested, a single dose of thalidomide could kill babies in the womb, degrade their organs, or lop off limbs." By 1960 the international medical community was circulating research papers showing a connection between thalidomide and these children.

At the same time, the American pharmaceutical company Richardson-Merrell (now Sanofi)† applied to the FDA for approval to sell thalidomide in the United States. Richardson-Merrell had already manufactured 10 million pills before applying for the FDA's approval, and their method for evaluating its safety was to distribute 2.5 million pills to physicians to give to their patients and see how they did. Patients and many of the physicians had no idea that Kevadon, as Richardson-Merrell labelled thalidomide, was not yet approved in the United States. At the FDA, Frances Oldham Kelsey and her team – Lee Geismar, chemist, and Jiro Oyama, pharmacologist

* Other physiological malformations, such as extra fingers or toes, partial or complete deafness or blindness, and injury to internal organs often resulting in death shortly after birth, also occurred.
† Although Richardson-Merrell passed through several acquisitions to become part of Sanofi, I am using the company's name at the time to avoid confusion with the present company. Sanofi did not exist during this time.

– rejected the application because of an absence of data on the medication's safety. She and the company went back and forth for months – Merrell submitted gibberish and meaningless numbers, Kelsey demanded real data and rejected the application, Merrell attacked Kelsey and submitted more gibberish and meaningless numbers* – until autumn 1961, when the research evidence of thalidomide's effects on adults and foetuses become impossible to ignore or deny. Chemie Grünenthal and the other companies took it off the market, even Merrell in March 1962. Around the world, thousands of adults had suffered irreparable nerve damage, and thousands of pregnancies had miscarried, ended in stillbirth, or resulted in infants with severe malformations and impairments. In the United States, the numbers were far lower thanks to the FDA's refusal to approve the drug.

The thalidomide disaster is important to this study for two main reasons: what happened, and how it is recounted and therefore represented. Chemie Grünenthal produced a substance through chemical experimentation and then went looking for a medical application, which was standard practice up to that time. That practice should sound familiar: it is what the chemists and chemical physicians from Chapters 4 and 6 had been doing. Physicians' ready acceptance of the drug makers' claims can also be traced to the Scientific Revolution, which built into Western medicine the concept that drugs are made from secret formulae in laboratories, places where ordinary people cannot go. The circulation of research papers establishing the connection between thalidomide and birth defects, stillbirths, and

---

* Kelsey's husband, Fremont Ellis Kelsey, PhD, called parts of the application "an interesting collection of meaningless pseudoscientific jargon apparently intended to impress chemically unsophisticated readers".

miscarriage was also put in place by the principles and practices of the Scientific Revolution, and the circulation of those
findings did exactly what it was supposed to do. Other aspects
of the crisis are the result of the New Science's tying together of
knowledge and gain: science used to make a profitable object
and to justify or promote that object, and purveyors of medication working with prescribers of medication for their mutual
profit. When science's function to provide benefits conflicted
with its function to provide profits, in the case of thalidomide
it was the former that gave way until 1960. Greed was not the
only force that enabled the thalidomide disaster. Other values
and practices that were built into science (and its offspring,
medicine) during the elimination of domestic medicine and its
replacement with modern or for-profit medicine also made it
possible.

In the United States, the story of thalidomide is recounted
as a David v. Goliath story, a tale of a heroic novice who triumphed over an evil empire (Evans called it the "imperial
power of American business"). Kelsey was indeed courageous
and unyielding in the face of tremendous pressure. She was also
a woman in her forties who had dealt with plenty of sexism and
who was rightly confident in her knowledge and skills. She had
a doctorate in pharmacology from one of the finest research
institutions in the United States (the University of Chicago),
which she had earned under the supervision of one of the most
distinguished pharmacologists of the day, E. M. K. Geiling.
Additionally, she had an MD from the University of Chicago
– which she earned while married and bearing and raising two
children – and had practised medicine for several years. She
kept up with the latest research and corresponded with others
in the field. She held to the principles of the New Science and
the New Medicine: observing, experimenting, analysing the

data, and drawing a conclusion from the data, all in a way that someone else could replicate, and she did it for the benefit of humanity. She was, in fact, a woman extremely well qualified for her role and consequently imbued with the power of knowledge and expertise.

Nor does it diminish her achievement to point out that she did not do it alone. The image of the individual genius in the laboratory is not how science (or medicine) is done. Kelsey had a team precisely because each one brought their own expertise in a relevant field to the drug evaluation process. She made the final decision, but their input shaped that decision. Kelsey also sought the expertise of pharmacologist Fremont Ellis Kelsey (her spouse, as it happened), who worked at the National Institutes of Health (NIH) and concurred with her analysis of Richardson-Merrell's application. The British physicians and apothecaries of the early and mid-eighteenth century would have agreed with Kelsey that Richardson-Merrell's claims about their "super drug" were dubious at best. Like Stephen Hartley (but not, unfortunately, Joseph Clutton), she had the support of colleagues and powerful institutions. When F. Joseph Murray, the executive at Richardson-Merrell in charge of the drug, complained to her boss, Ralph Smith, Smith told him that Kelsey had the full support and weight of the FDA behind her. She did exactly what she was supposed to do as a scientist and physician, and the scientific community (which includes the medical community) did exactly what it was supposed to do.

But Frances Oldham Kelsey also defied three centuries of science. She did not just want data establishing the medication's safety for patients. She had done considerable research in teratogens, substances that cross the placenta from mother to foetus, and studied embryology. She wanted data proving that thalidomide was safe for pregnant patients, and she wanted data proving that it was safe for their foetus. In one way, this was good science, but in another way it was deviant science. Kelsey rejected the idea of a "universal subject", a human being whose responses to medications (or anything) were shared by all other human beings. The "universal subject" came out of the Scientific Revolution, which pushed women out of sight as much as possible when it redefined and took over medication. Difference, women's bodies, and women's experiences were all erased as much as possible. Kelsey brought women's bodies back into everyone's awareness and into the centre of the discussion of a medication that was – oh, right – given to women, and for a – oh, right again – specifically female experience. She and Dr Helen Taussig, internationally renowned as one of the pioneers of paediatric cardiology, ensured that the development of the foetus and the children who were born was also recognized and accounted for. Kelsey made women's bodies, women's experience, and the foetus they were making central to the analyses of thalidomide – not only visible but also vital.

There is some truth to the adage that the more things change, the more they stay the same. After thalidomide, governments around the world, including the United States, revamped their criteria and processes for medication approval. Drugs must undergo rigorous testing. Patients must be informed if the drug they are taking is experimental and must be given the choice whether to participate in clinical trials – what is called informed consent. Nevertheless, the regulation of medication inevitably continues to conflict with the commodification of medication. Agencies such as the FDA and the European Medicines Agency have "fast tracks", routes that certain drugs can take if they are so urgently needed that final rounds of testing can be deferred until after their release. (There are also emergency authorization protocols for medications such as the initial coronavirus disease 2019 vaccines.) The example of aducanumab, a drug proposed for treating patients with Alzheimer's disease, is the reverse of thalidomide, and reinforces the old lesson that regulation requires both science and the will to heed it.

Health Canada, the European Medicines Agency, and the United Kingdom's Medicines and Healthcare products [*sic*] Regulatory Agency all rejected Biogen's application for aducanumab because the data about its efficacy were inconclusive and showed the possibility of life-threatening and fatal side effects. The FDA, however, ignored the recommendation of its own review committee to reject the application. Ten out of eleven members of the committee voted against it (the eleventh voted "unsure") but aducanumab was approved, and for a larger patient pool than it was designed for (although this was changed later). Three committee members resigned and the subsequent investigation by the Office of the Inspector General found that the FDA used (deliberately or de facto) its "accelerated approval pathway" to allow pharmaceutical companies to avoid completing their safety and efficacy testing. The FDA had gotten too close to the corporations that it was supposed to be policing and regulating. All the Inspector General's Office had to do was read *Forbes*: Robert Hart reported that Roche was "the latest pharmaceutical company to court the regulator in the wake of following its controversial fast-track approval of Biogen's Aduhelm [aducanumab]". Regulators are not supposed to be woo-able, but the FDA had signalled that it was.

The gender imbalance in science, technology, engineering, and medicine (frequently shortened to "STEM") has been increasingly publicized and addressed over the last few decades, but unfortunately it is only one consequence of the Scientific Revolution's construction of gender to define and establish science and medicine. The last pair of medications, fezolinetant and sildenafil, demonstrate other consequences of removing and barring women from the medication marketplace. Fezolinetant was created by Astellas Pharma to reduce moderate-to-severe hot flushes (sometimes called hot flashes) in perimenopausal women. It is the only drug of its kind, unless Bayer wins approval for elinzanetant. In 2023, fezolinetant was approved by the FDA and was undergoing review by the European Union, the United Kingdom, Switzerland, and Australia.

Perimenopause, which can last up to a decade, is the stage during which menstruation slowly ceases due to a lowering of hormones, particularly oestrogen. Menopause is the stage after menstruation has stopped; loss of oestrogen causes a seemingly infinite number of other health problems during this stage. Up to 80 per cent of perimenopausal women experience hot flushes, which can interrupt ordinary work and daily life. According to one study, between 10 and 20 per cent of women rated their hot flushes "near intolerable". Hot flushes also have a cascading effect, disrupting sleep and concentration, which in turn affect physiological and mental health, memory, social interactions, and the ability to perform paid and unpaid work.

Until recently, however, the question of economic productivity and menopause was characterized by researchers as a "quality of life" issue, as if the hours of work lost have nothing to do with paying for food and shelter, or with an employer receiving the employee's best work. In reality, the financial impact of debilitating menopause symptoms hits individuals and families as well as corporations and national economies. By one calculation, the US economy loses almost 2 billion dollars per year. More than half of the world's population is biologically female and every single one who survives long enough – not at all a foregone conclusion in many places – will experience menopause. That is a lot of people, a lot of families, a lot of local and national and global economic impact. So it is a good thing that researchers have been working so long and assiduously to treat erectile dysfunction (ED).

The first medication to treat ED, sildenafil (Viagra), was approved in the United States and Europe in 1998, a quarter of a century before fezolinetant. As with thalidomide and many other medications, sildenafil's effectiveness at treating this condition was an unexpected but very welcome discovery. Since then at least two other prescription medications for ED have hit the market, and generic forms are also available. Medications like sildenafil treat ED, a condition in which an adult man cannot get an erection or maintain it sufficiently for sexual activity with or without a partner, by enabling the patient to get and maintain one. ED is caused by conditions such as diabetes, certain kinds of injury or surgery or treatment, and ageing. ED itself does not prevent someone from working, but a root cause, such as a heart condition or cancer treatment, or a psychological response to ED, such as depression, can have that effect.

Without minimizing the impact of ED on those suffering from it, the choice by researchers and corporations to address it before giving attention to any of the effects of the perimenopause or menopause makes little fiscal or pharmacological sense. Economically speaking, more people are affected by menopause than by ED, so the market is greater. Medically, hot flushes themselves impair functioning, so medication for it addresses the actual health problem. Because ED is often a symptom of a serious underlying ailment that requires treatment, medication for it does not address the real health problem.

Furthermore, one could argue that the psychological distress caused by ED is itself the product of definitions of masculinity.

If so, then addressing those definitions as well as attitudes towards ageing would be more helpful and wide-reaching for treating the psychological response to ED than restoring the ability to maintain an erection long enough for penetration. The word "dysfunction", for example, suggests that the penis is not working properly, but if inability to maintain an erection for a certain period of time is a natural effect of ageing, then perhaps the condition when caused by ageing deserves a different name. Menopause is not "menses dysfunction" after all. But to return to the pharmaceutical makers and distributors – despite the size, scope, and permanence of the market and the impact of a medication for hot flushes, researchers and corporations invested in ED, and more than once. That repeated decision suggests that it is not the condition so much as who suffers from it that guided the decision.

The amount and relative timing of medications for ED and for the hot flushes of menopause are more than the artefacts of sexism. They are also the products of a particular form of sexism rooted in the appropriation of medication by male proponents of the New Science. "Medicine carries the burden of its own troubling history," Elinor Cleghorn writes. "The history of medicine, of illness, is every bit as social and cultural as it is scientific." As science's revolutionaries pushed women out of certain figurative and literal spaces and into smaller ones, women became less and less figuratively and literally visible. Then and now, problems of access, patient credibility, and quality of treatment are compounded by race, ethnicity, class, and sexual identity. In 2021, the overall maternal mortality rate in the United States was 32.9 deaths per 100,000 live births (a 40 per cent increase from 2020), by far the worst of the industrialized countries, which average 12 deaths per 100,000 live births. The maternal mortality rate for American Black women, however, was 69.9 deaths per 100,000 live births. The degree of disparity in the United Kingdom was comparable.

Frances Oldham Kelsey's demand for studies assessing the risk of thalidomide to women and pregnant women shocked Richardson-Merrell's executives because nobody had thought about the female body as its own being – a blindness that, as extensive research shows, continues in clinical trials in the twenty-first century. It is not a coincidence that "Veozah" sounds a lot like "Viagra": the name capitalizes on the associations raised by the sounds (V-vowel-vowel-hard consonant-ah) and rhythm, and also suggests an equivalence between the drugs. Nevertheless, Veozah's arrival on the market twenty-five years after Viagra is a signal of just where women's physiological debility rates in comparison with men's. Anyone who is not the universal subject can be a deviation or a deviant, but they cannot be someone with integrity of being. The simultaneous subordination and subsumption of the non-universal subject is dangerous if not potentially fatal for such an individual or demographic.

The commodification of medication during the seventeenth and eighteenth centuries by proponents of the Scientific Revolution was accomplished by the establishment of ideas, values, principles, and systems that displaced domestic medicine and its values and systems. What the Scientific Revolution brought, and the medication marketplace that came out of it, has left Western medicine and the global pharmaceutical system with a deeply problematic legacy. Domestic medicine's principles and ethos are not gone, however, and on occasion over the last century, domestic medicine has been used in conjunction with modern science and medicine to very positive effect.

Consider iodized salt. Humans need both sodium and iodine to function. Sodium makes nerves work and helps maintain the right amount of blood plasma, while iodine is necessary for the thyroid to make hormones that regulate metabolism, and therefore affect the growth and functioning of the entire body. Iodine deficiency can make the thyroid gland enlarge and bulge out of the neck, which is called goitre, and causes neurocognitive and neurological damage. Iodine deficiency in utero can lead to stunted growth and weight gain, muteness, deafness, impaired ability to control limbs, impaired motor skills, and even thyroid cancer. As one research team put it, "Fetal development highly relies on thyroid and iodine metabolism and can be compromised if they malfunction."

By far the best delivery method for salt and iodine is eating things with salt and iodine. Salt is pretty easy to find in nature; iodine, not so much. It is only available naturally in edibles that grow in iodine-rich environments, like the ocean or soil near the sea. Take a look at a map of the earth: many people do not have ready access to those foods. At the turn of the twentieth century, in some areas of landlocked Switzerland, 75–80 per cent of schoolchildren had an enlarged goitre. In the "Goiter Belt", a band of territory stretching across the top of the United States from upper New York State to the North-west, 25–75 per cent of schoolchildren had it. Interest in addressing iodine deficiency intensified around 1914, at the outbreak of First World War. In the United States, the number of men disqualified from military service because of goitre was so high that it affected military strength. As Hugh Chamberlen had pointed out more than two centuries earlier, "To preserve Health and save Lives, is always a Public Good, but more especially in time of War."

Domestic medicine provided a solution. At the time, medication and food were sufficiently separate categories that despite knowing about iodine in food, some public health officials proposed administering iodine drops to children. Domestic medicine did not draw such a strong distinction, however, and also held the view that eating and drinking is a form of healthcare. Between 1915 and 1918, Swiss and American physicians began exchanging information and reading each other's research on iodine additives in the school-age population. The results were very promising. Salt was a common household item, a basic ingestible, and iodizing it worked extremely well. Iodized salt is also cheap, easy to produce, extremely effective, and voluntarily consumed by people without thinking too much about it. No "Time to take my iodine", just "What's for dinner?" Once scientists, physicians, public health officials, and the executives of salt companies abandoned that firm distinction between medication and food, they had a solution to widespread iodine deficiency: iodized salt, which everyone could keep on the kitchen table and eat regularly. Is iodized salt food or medication? Yes.

What happened after American and Swiss healthcare providers latched onto iodizing salt, however, is a tale of divergence. Switzerland forged ahead with a national salt iodization programme in 1919 and implemented it in 1922. By the early twenty-first century, iodized salt was in the kitchen of more than 94 per cent of Swiss households. Countries around the world followed suit. Well over 100 now require table and cooking salt to be iodized. In 1990, the World Summit for Children made "Universal access to iodized salt by 1995" one of its goals. At the time, 130 countries grappled with iodine deficiency. By 2016, that number had dropped to nineteen, with thirty-six countries in Asia alone implementing or developing legislation to provide iodized salt for their citizens. On the whole, citizens of countries with iodized table and cooking salt suffer much less iodine deficiency and fewer physiological ills, they bear healthier children, and they have higher IQs than their forebears in the days before iodized salt.

Additionally, continuing to ignore the distinction between food and medication, scientists have continued to find comestibles to iodize. The idea to iodize salt came from observing one culture using naturally iodized salt, but the idea to iodize other items comes from a willingness to continue blurring that boundary. Scientists and physicians accepted the kitchen as the key space in which cures and prevention were to be worked. These elements from domestic medication – the relaxed distinction between food and medication, the community orientation, the kitchen at the centre of health – combined with scientific principles and methods to create an effective solution to a global problem. Governmental commitment to the community, to public health, is vital. The closer that a government gets to mandating that food-grade salt be iodized, the less iodine deficiency there is in the populace.

The United States went about it very differently. There was no national or state legislation. An attempt at the federal level in 1948 to ensure the availability of iodized salt to all consumers failed. Instead, dealing with iodine deficiency through iodized salt began with the corporate sector. Michigan physician and professor of paediatrics Dr David Murray Cowie persuaded the salt corporations to make iodized salt for the Michigan market, and it arrived on grocery store shelves in 1924. "Incredibly," writes Howard Markel, "the salt manufacturers voluntarily produced every box; no laws or regulations were required." The introduction of iodized salt to consumers was so successful that the Morton Salt Company took its product national within months, proving that public health and corporate interest are not inevitably in conflict.

Consistent with findings that greater government investment

in getting iodine into its citizens produces less iodine deficiency in its citizens, however, iodine deficiency is on the rise
in the United States. In 2022, the research team of Adrienne
Hatch-McChesney and Harris R. Lieberman reported that in
2011–2012, almost 40 per cent of Americans were iodine deficient; a more recent study showed that 23 per cent of pregnant
women in Michigan were iodine deficient (ironic considering
that Michigan was where iodized salt first became available).
In addition, as of 2015, iodine deficiency was on the rise in the
American military. Other industrialized countries, particularly
in Europe, have the same problem.

Iodine deficiency is most effectively reduced when boundaries
between food and medication are relaxed, and its role as a commodity is de-emphasized – when it is treated as a national issue
requiring public-sector action. This is not to advocate medical-
ized food, "wonder foods" that pack a full day's vitamins and
minerals into a single processed bar, for instance, or eating a
certain fruit or vegetable because it supposedly enhances brain
functioning or calorie burning. Treating comestibles as tools for
engineering a physiological effect (More energy! Better sleep!
Stronger memory!) where there is no actual medical need such
as iodine deficiency creates a medical problem where there is
none. It also perpetuates the ideas of the body as a mechanism
to be manipulated and health as a state to be restored. That
does not mean that corporate and public interests cannot coincide. On the contrary, the initial success of iodized salt in the
1920s in the United States or India's recruitment in the 1980s
of private salt companies to assist the government's iodization plants show that public health and financial gain can be
arranged to work in concert, and not solely to produce citizens
fit for war – "food for powder", as Shakespeare's Falstaff says.

Domestic medicine's view of health and treatment as

community concerns has also proved powerfully effective when applied in health crises. The development of treatments for HIV/ AIDS in the 1980s and 1990s in the United States illustrates this point. Obeying both Francis Bacon's contention in 1605 that methods of inquiry have to be designed to nullify human biases, and the stricter regulations put in place after thalidomide in 1962, drug trial protocols at agencies around the world became thorough, cautious, and consequently lengthy. In the case of AIDS in the United States, those protocols put the development of treatments on a far slower schedule than the one imposed by the virus on the people it infected. Although researchers at the National Institute of Allergy and Infectious Diseases (NIAID), a division of the NIH, devised a new procedure for testing medications that they considered speedy, it did not seem so when compared with the virus's timetable. To infected people, most of whom were gay and all of whom were dying with appalling speed and in horrific numbers, these protocols were too removed from human concerns and far too slow. "When there were no treatments," Sarah Schulman wrote, "everyone was going to die." The AIDS Coalition To Unleash Power (ACT UP) began publicly and aggressively confronting the scientists (also the public and the federal government) with this charge. The research community initially replied with responses first drafted centuries earlier: only people with special training can understand this stuff, it is too hard for ordinary people, you are too emotionally invested, you are irrational, you cannot do science if you have something at stake in the results, leave it to the experts.

However, "Regulators must find a way to allow access to potentially life-extending therapies," an editorial in the *Lancet* stated in 2021, "while both scientific rigour and the safety of patients are maintained." Everything changed once the researchers and the activists began speaking to each other. Unlike thalidomide, AIDS medications were not approved without thorough testing, but access to experimental treatments expanded. One early, major development in the United States was NIAID's creation of a "parallel track program" that gave AIDS patients access to experimental drugs without having to be in a drug trial. It was a purely voluntary programme, and although there was no guarantee that the drug would work, there was also no chance of receiving a placebo. Unlike aducanumab (Aduhelm), medications for AIDS were approved at the recommendation of the review committees.

Furthermore, although ACT UP was predominantly white and male, its methods and "patient-centered politics, learned from feminism", as Schulman explained, "achieve[d] transformational victories that improve the lives of women, people of color, and poor people". Testing protocols became much more inclusive. HIV-positive people joined research programme supervisory committees. It turned out that the specialists with doctorates were not the only people who could understand retrovirals and science-y stuff. The many voices of those affected by AIDS also convinced officials to expand the definition of AIDS to include symptoms that women but not men experienced, such as acute pelvic inflammatory disease and constant yeast infections, thus recognizing the female body and its difference from the male body. Incorporating affected people into medical research, factoring in human issues such as the rate at which an illness hastens the patient towards death, and working with and within the community have become the norm in many areas of medical and scientific research, including at the FDA and the European Medicines Agency.

The *Lancet*'s editorial made a significant mistake, however, in drawing an analogy between the FDA with aducanumab

(Aduhelm) and the pressure from advocacy groups on Anthony Fauci, then Director of NIAID, to approve untested or under-tested treatments. "At the time," the editorial explains, "he *encouraged changes to the approval process* both to expedite approval and to allow greater patient participation in clinical trial design and approval. For aducanumab and other treatments, regulators must find a way *to allow access*" (added emphasis). Fauci and NIAID did not respond to external pressure by approving an insufficiently studied medication. Rather, Fauci agreed to include more stakeholders in the process, to make the process less opaque and less paternalistic.

A central element of domestic medicine, working with and within the community, has proven vital to other successes. The eradication of smallpox by 1980 would have been impossible without community involvement. The World Health Organization (WHO) launched its global campaign to eradicate smallpox in 1959, but it made almost no headway until 1967, when its approach changed from swooping in with orders and vaccines to partnering with the local community. While the WHO was failing, however, a number of countries and regions established their own, more successful smallpox eradication programmes, forming national, regional, and global webs of people (politicians, tribal leaders, medical professionals, international health officials, local health officials, people in every walk of life, the old, schoolchildren – everyone) working for the same goal.

The WHO has learned a lesson, however. For another scourge, river blindness, the organization now recommends that each country devise its own programme for eradicating it. River blindness, medically known as onchocerciasis, is produced by a parasite that causes skin disease, eye injury, and blindness. A number of countries working to eradicate it formed partnerships with the Carter Center, a non-profit organization founded by Nobel laureate and former US President Jimmy Carter and former first lady Roslyn Carter. Since 1997, one of the Carter Center's goals is the eradication of river blindness. Merck, which makes ivermectin (Mectizan), the medication that kills the parasites, has also joined the partnership: the company donates Mectizan to the Carter Center's programme. These partnerships between national ministries of health, the Carter Center, and Merck are based in the country where the disease is, and the local experts and community leaders take the lead. It is an extremely successful strategy. River blindness has been entirely eliminated in four countries, almost eliminated in two more, and its transmission has been interrupted in swathes of Africa.

"Local" or "community" can refer to national efforts or partnerships, but the term also applies to the very smallest community units. In Malawi, a programme to eradicate sleeping sickness uses theatre performances at community soccer fields and open spaces to teach people about it. The performance was developed collaboratively by parasitologists, epidemiologists, survivors, performing artists, and public health workers from Scotland and Malawi, and involves audience participation as a group. At the end of every performance, the energized, engaged audience has the opportunity to ask questions of the troupe, and very early data indicate that more people took the recommended precautions after the show than before it.

Bangladesh employs an army of women as local health officers (called "shasthya kormi") responsible for treating malaria, another parasitic disease. Everyone in the community knows the woman they can go to if they are sick, and she has the equipment to test for malaria; the training to use it, diagnose, and treat malaria; and the authority to distribute artemisinin, the medication. The programme is a tremendous success, but typical of science's emphasizing technology and medication and erasing women and the domestic, discussions and descriptions of the programme often overlook the significance of the shasthya kormi and emphasize the medication. According to one reporter, "Between 2008 and 2020, malaria cases in Bangladesh plummeted by 93%, thanks largely to this drug." Artemisinin obviously was vital to the precipitous drop in cases, but even miracle drugs cannot walk from the medicine cabinet to the patient on their own. That requires the trained, dedicated women who look after their communities, monitor illness, and possess the medical knowledge and respect to treat it.

Artemisinin is derived from a plant that ought to sound familiar to readers of Part One: wormwood. One species of wormwood, *Artemisia absinthium*, is indigenous to Europe, including Great Britain. That is the wormwood that James Douglas asked the laboratory on "Cheesewell" Street to analyse (see Chapter 6). Another species of wormwood, *Artemisia annua*, is indigenous to China and is the source of quinine. The earliest recorded appearance of wormwood in traditional Chinese materia medica is the fourth century CE, but the search for wormwood's active medicinal compound began in the 1960s in China and concluded with the identification of artemisinin in 1979. Artemisinin and its laboratory-modified versions have been the world's front-line treatment for malaria ever since, with scientists racing to manipulate the original medicinal compound to keep ahead of the development of artemisinin-resistant strains. Laboratory-synthesized versions of artemisinin are vital for another reason: the plant does not grow just anywhere, and a large amount of wormwood is needed to obtain a useful quantity of the compound.

Quinine and its offspring antimalarials constitute one

admittedly compelling example of ethnomedicine, a combination of anthropology and medicine that studies traditional medications in cultures around the world. In the case of quinine, researchers in China scoured their own medical tradition for possible malaria treatments. The term for this is "bioprospecting" and it includes analysing recipes from ancient herbals and materia medica chests in search of medicinal compounds. Scientific analysis and testing have established that traditional treatments such as thyme, saffron, yarrow, sage, and matalafi leaves, used in traditional Samoan medicine for inflammations, really do have the effect that different medicine traditions say they do, which makes these organics potential medicinal bases for pharmaceutical companies. As Seeseei Molimau-Samasoni's research team explained: "Compounds from natural resources are reliable lead templates of new pharmaceuticals, having persisted through evolutionary selection to control fundamental molecular pathways," and because "Medicinal plants in effect have been trialed for activity through centuries of ethnobotanical use [...] traditional medicines [are] an attractive yet challenging source for further investigation." In other words, plants are a great foundation for building curative molecular compounds, and medications used by local healers are a great starting place because they would not be in use if they did not have medicinal properties. Bioprospecting is not about proving that traditional medications work; it starts from that position, then uses scientific methods to identify the (hopefully) replicable or modifiable medicinal compound in the plant.

Bioprospecting is not new, and it goes hand in hand with what intellectual property rights advocates call "biopiracy": taking knowledge from local healers and turning it into a valuable commodity without sharing the profits. The Madagascar or rosy periwinkle, for example, produces two powerful anticancer chemicals, vinblastine and vincristine. Vinblastine's spectacular profitability is not shared with any of the several countries that credibly claim to have provided knowledge of the periwinkle's medicinal properties. It should be more difficult to commit biopiracy since 1992, when the Convention on Biological Diversity (CBD) was signed at the United Nations meeting in Brazil, although the CBD has no enforcement mechanism and is controversial in some quarters. Nevertheless, the global conversation around organic medications has become much more interested in collaboration and consortia to find natural materials and develop medicines from them. Without idealizing anyone or anything, some organizations, governments, and individuals seem to be finding ways to balance the drive for profit that commodified medication with other considerations: partnership, equality, fairness, sharing knowledge, even preserving resources for the future. For example, Molimau-Samasoni's team declared their principles in their published findings:

> Working with traditional healers via an ethical, data sovereignty-driven collaboration led by indigenous Samoan researchers, we elucidate the chemical biology of the poorly understood but often-used Samoan traditional medicine "matalafi," the homogenate of *Psychotria insularum* leaves commonly used to treat inflammation-associated illnesses. Our approach unifies genomics, metabolomics, analytical biochemistry, immunology, and traditional knowledge to delineate the mode of action of the traditional medicine rather than by the more common reductionist approach of first purifying the bioactive principles, which can be used to better understand the ethnobotany of traditional medicine.

Developments in bioprospecting suggest that the values of domestic medicine can work effectively with modern science without imperilling anyone's profit or safety. It requires lateral relationships, sharing knowledge and skills, instead of vertical relationships, handing down orders and information.

Acknowledging the herbaceous origin of a medication – or of a huge number of medications – also takes the laboratory and chemistry off their pedestals. Biochemists and botanists are both necessary to pharmacological creation. Admitting partnerships changes the perception of science and medicine, as well as perceptions of the creation of a commodity, and of the roles of humans and of nature. If pharmaceutical companies were willing to admit which and how many of their pharmaceuticals began as a leaf, root, berry, or stem, for instance, their laboratories might have to share centre stage with nature. On the other hand, admitting how much they rely on what is growing out there would call attention to nature, and put a human as well as a financial value on conservation. Think of all the treatments and cures that will never happen if the moss or tree or flower or grass that they come from becomes extinct. Imagine what climate change – what just the Amazon rainforest – would look like if Big Pharma fully and publicly used its muscle to preserve and protect nature.

Put your finger on any nation on the map and the chances are good that some of its population is suffering, perhaps dying, because there is no money to get them the medication that they need. But modern medicine and modern science do not require the commodification of medication. It is a concept that was invented and introduced, and that eventually became the norm not because its invention or acceptance was inevitable, but because of the decisions and actions of people. For the same reason, medication as a commodity and the system that protects (and hides) its commodity status are not immutable. Overhaul, modify, or keep the medication system as is – those actions are all choices. Not doing anything is as much a decision as making change. Of course, whatever you do next, dear reader, is up to you, but whether you act to change or to preserve things as they are, you have made a decision (although not an irreversible one).

The period between 1650 and 1740 in Great Britain saw the replacement by the Scientific Revolution's adherents of an old system of values and treatments with new, "revolutionary" ones. Medication was redefined into a narrower category by its distinction from food, it was re-gendered into a masculine item and a masculine sphere, and it was transformed from a household item and a right into a commercial item and a commodity with limited access. Profit competed with compassion and care, and often triumphed, even when it meant other peoples' injury or death. In the centuries since, there have been some important changes: science's medications have become truly superior to whatever seventeenth-century women or apothecaries were making, and scientists have learned to modify and improve on the original curative elements in a medicinal substance. Moments of crisis have put pressure on the system built by the New Science's revolutionaries and have shown how domestic medicine's irrelevance has been greatly exaggerated. Recovering women in history, and in the history of medicine and science, also recovers values and practices that once upon a time served humanity well.

Bibliography

Of her book *The Restless Republic*, historian Anna Keay wrote that "Nothing would make its author happier than that it inspires the reader to search out other books on this period." The same goes for the author of this book, with the added desideratum that the reader should also be inspired to seek out primary materials from the period. Accordingly, this bibliography is divided into several sections: digital collections available to the general public, framework texts that shaped the entire project, and chapter bibliographies of materials specific to that topic.

DIGITAL COLLECTIONS

There is a wealth of digitized materials on the history of science and the history of medicine. Much of it can be viewed with only an Internet connection. The most famous libraries, such as the British Library and the Library of Congress, have digitized many different collections, as have a number of academic institutions such as the University of Cambridge, the University of Oxford, and Yale University. Specialized libraries, such as the Wellcome Collection, the National Library of Medicine (United States), the New York Academy of Medicine, and the Dibner Library of the History of Science and Technology of the Smithsonian Libraries and Archives have particularly extensive digital collections, as do the libraries at institutions such as the Royal College of Physicians and the Royal Society.

Many libraries also have research databases containing scanned or digitized printed materials, such as Early English Books Online, Eighteenth-Century Collections Online, and the Seventeenth and Eighteenth Century Burney Newspapers Collection. The *Oxford Dictionary of National Biography* is invaluable; I have used it for biographical information throughout my research. However, access to these databases can be restricted to library card holders. Fortunately, there are a number of open-access databases. Of note are Munk's Roll, a list of Fellows of the RCP since its inception, on the RCP website; the Hartlib Papers at the University of Sheffield; the Early Modern Practitioners project at the University of Exeter; and British History Online.

FRAMEWORK MATERIALS

This category includes resources that I used throughout *The Apothecary's Wife*, not only for specific points of information or insight but also to deepen and broaden my understanding of the many aspects of this subject. Accordingly, they will not be listed in chapter bibliographies.

Despite the plenitude of digitized materials, the vast majority of seventeenth-century and eighteenth-century manuscripts and rare books exist only in material form and must be visited in person. In many places, especially county record offices, permission to see rare books and manuscripts comes with membership. The most important sites for *The Apothecary's Wife* have been the British Library, particularly for the Sloane Manuscripts, Brockman Papers, Blenheim Papers, and Miscellaneous Papers relating to the Twysden family; Chawton House Library; the London Metropolitan Archives; the Royal College of Physicians heritage library and archives; the Royal Society; the Society of Antiquaries; Lambeth Palace Library for the Fairhurst Papers, Shrewsbury letters, and Miscellaneous Papers of John Selden and Sir Matthew Hale; the University of Aberdeen's Special Collections; the University of Glasgow's Hunterian Collection; and the Wellcome Collection. I am also indebted to the Devon Record Office, the Sheffield City Archives for the Arundel Castle Manuscripts, the Bedfordshire Record Office, the Warwickshire Record Office, and the Wiltshire and Swindon Record Office for access to their holdings.

It should not surprise anyone to read here that many histories of science and histories of medicine are deficient – to be kind – when it comes to gender, race, ethnicity, class, and imperialism. Fortunately, that situation is changing at the academic level and more slowly for general audiences. (Even books that proclaim their even-handedness manage to disappoint; their index – non-fiction's X-ray – gives them away.) Jack Turner's witty, highly readable *Spice: The History of a Temptation* (2004) is a good starting place. A similar problem attends the history of medical professions in the British Isles. For a long time, histories of apothecaries and physicians were not only all male but also focused on London. These works are useful for getting a sense of chronology, events, and people, and they have informed this book throughout. Penelope Hunting's *A History of the Society of Apothecaries* (1998) is the most recent, concise, and reliable account. I recommend anything by her, actually. Hunting's study was preceded by E. Ashworth Underwood et al., *A History of the Worshipful Society of Apothecaries*, vol. I, 1617–1815 (1963) and W.S.C. Copeman, *The Worshipful Society of Apothecaries of London: A History, 1617-1967* (1967), which are products of their time. Shorter works that focus on the origins of the apothecary include Juanita G.L. Burnby, "The Apothecary as Progenitor", *Medical History* 27, n. S3 (1983), pp. 24–61; John A. Hunt, "The Evolution of Pharmacy in Britain (1428–1913)", *Pharmacy in History* 48, n. 1 (2006), pp. 35–40; Penelope J. Corfield's invaluable "From Poison Peddlers to Civic Worthies: The Reputation of the Apothecaries in Georgian England", *Social History of Medicine* 22, n. 1 (2009), pp. 1–21, and C.J.S. Thompson, "The Apothecary in England from the Thirteenth to the Close of the Sixteenth Century", *The History of Medicine* 8 (1915), pp. 36–44 (old but still useful). The most recent history of London physicians is *500 Years of the Royal College of Physicians*, edited by Linda Luxon and Simon Shorvon (2018). Their work follows the path laid down by George Clark's two-volume *History of the Royal College of Physicians of London* (1964).

For information about practitioners outside London during the seventeenth century and eighteenth century, read almost anything written by Margaret Pelling, such as "Barber-Surgeons' Guilds and Ordinances in Early Modern British Towns – the Story so Far", Working Paper n. 1, *Early Modern Practitioners Working Papers*. Additional informative publications are Alun Withey's "'Persons That Live Remote from London': Apothecaries and the Medical Marketplace in Seventeenth- and Eighteenth-Century Wales", *Bulletin of the History of Medicine* 85, n. 2 (2011); David Harley's "'Bred up in the Study of that Faculty': Licensed

Physicians in North-West England, 1660–1760", *Medical History* 38 (1994), pp. 398–420; Steven King and Alan Weaver, "Lives in many Hands: The Medical Landscape in Lancashire, 1700–1820", *Medical History* 45 (2000), pp. 173–200; and G.H. "Apothecaries in Early Modern Edinburgh", *Pharmacy in History* 37, n. 3 (1995), pp. 135–36.

Fortunately, an efflorescence of research in history, its subfields, and in literature is returning women to the historical record. Early groundbreaking works – *The Mind Has No Sex? Women in the Origins of Modern Science* (1989) and *Nature's Body: Gender in the Making of Modern Science* (1993) by Londa Schiebinger; *Women, Science and Medicine 1500–1700*, edited by Lynette Hunter and Sarah Hutton (1997); and *Men, Women, and the Birthing of Modern Science*, edited by Judith P. Zinsser (2005) – opened the field significantly. Elizabeth Potter's demonstration in *Gender and Boyle's Law of Gases* (2001) that definitions of masculinity were integrated into emerging science laid the foundation for works such as Elinor Cleghorn's *Unwell Women: Misdiagnosis and Myth in a Man-Made World* (2022), Anushay Hossain's *The Pain Gap* (2022), and Caroline Criado Perez's *Invisible Women: Data Bias in a World Designed for Men* (2021). Sheilagh Ogilvie's comprehensive *The European Guilds: An Economic Analysis* (2021) uncovers the reality of women in guilds. In addition to those mentioned in this part of the Bibliography, a few notable books include Lyn Bennet, *Rhetoric, Medicine, and the Woman Writer 1600–1700* (2018); *Inventing Maternity: Politics, Science, and Literature, 1650–1865*, edited by Susan C. Greenfield and Carol Barash (2015); *Women Philosophers from the Renaissance to the Enlightenment: New Studies*, edited by Ruth Habensgruber and Sarah Hutton (2021); *The New Science and Women's Literary Discourse: Prefiguring Frankenstein*, edited by Judy A. Hayden (2011); and my own explanation of how the Scientific Revolution and the invention of the novel fed each other, *Women, the Novel, and Natural Philosophy, 1660–1727* (2014).

The figure of the heroic individual appears frequently in histories, vide biography's popularity. There are stand-alone tomes on any number of science's revolutionaries, such as Hans Sloane (James Delburgo, *Collecting the World: Hans Sloane and the Origins of the British Museum* [2017]), Robert Hooke (Lisa Jardine, *Robert Hooke: The Curious Life of the Man Who Measured London* [2004]), and Nicholas Culpeper (Benjamin Woolley, *Heal Thyself: Nicholas Culpeper and the Seventeenth-Century Struggle to Bring Medicine to the People* [2004]). Readers interested in Vesalius and Paracelsus can find illuminating introductions in Stanley Finger's *Minds Behind the Brain: A History of the Pioneers and Their Discoveries* (2000); Joseph F. Borzelleca, "Paracelsus: Herald of Modern Toxicology", *Toxicological Sciences* 53 (2000), pp. 2–4; and Steven A. Edwards, "Paracelsus, the Man Who Brought Chemistry to Medicine", *American Association for the Advancement of Science* online (2012). The contrast between men and women in this regard is striking. For example, the most recent biography of Lady Mary Wortley Montagu, who brought inoculation to Europe from the Ottoman empire, persuaded the British medical community and the royal family of its efficacy, and consequently saved tens of thousands, probably hundreds of thousands of lives, came out in 2021 (*The Pioneering Life of Lady Mary Wortley Montagu* by Jo Willett); before that, there was Isobel Grundy's biography of her in 2004. In contrast, biographies of Edward Jenner, who created vaccination eighty years later, were published in 2020, 2022, and 2023. On the whole, there have been very few book-length biographies of seventeenth-century and eighteenth-century women involved in the Scientific Revolution between 1998, when Anna Battigelli published *Margaret Cavendish and the Exiles of the Mind*, and 2021, when Michelle DiMeo published her biography of Katherine Jones, Lady Ranelagh.

There are several texts that I recommend for getting a sense of what seventeenth-century and eighteenth-century life was like. To understand the buying power of money during the period, readers can rely on Robert Hume's article, "The Value of Money in Eighteenth-Century England: Incomes, Prices, Buying Power – and Some Problems in Cultural

Economics", *Huntington Library Quarterly* 77 n. 4 (2015), pp. 373–416. *London: Prints and Drawings before 1800* by Bernard Nurse (2017) is dazzling, with beautiful reproductions of some of the best maps of the seventeenth and eighteenth centuries, right down to the garden plots behind houses. David Cressy's *Birth, Marriage, and Death: Ritual, Religion, and the Life-Cycle in Tudor and Stuart England* (1997) is still one of the best accounts of lived life during that time. Sara Read's *Menstruation and the Female Body in Early Modern England* (2013) and Sarah Fox's *Giving Birth in Eighteenth-Century England* (2022) offer valuable insight into women's bodily experiences. *Coffers, Clysters, Comfrey and Coifs: The Lives of Our Seventeenth Century Ancestors* by Janet Few (2012) recounts daily life through the common objects that people used; it is very readable and informative, and not overly scholarly. *The Country House Kitchen, 1650–1900*, edited by Pamela A. Sambrook and Peter Brears (1996) takes the audience on a room-by-room tour of the development of domestic space dedicated to the procurement and processing of medication and foodstuffs. *Family and Business During the Industrial Revolution* by Hannah Barker (2017) provides a different perspective on daily life for women, men, and children. For recipe books, readers might begin with *Reading and Writing Recipe Books, 1550–1800*, edited by Michelle DiMeo and Sara Pennell (2013); *Recipes and Everyday Knowledge: Medicine, Science, and the Household in Early Modern England* by Elaine Leong (2018); and *Recipes for Thought: Knowledge and Taste in the Early Modern English Kitchen* by Wendy Wall (2016). Kristine Kowalchuk's edition *Preserving on Paper: Seventeenth-Century Englishwomen's Receipt Books* (2017) is an excellent place to begin reading the books themselves before tackling the convolutions of seventeenth-century handwriting in digitized manuscripts.

Where a specific item is significant to the chapter, I have identified it in the chapter bibliography. Otherwise, readers should consult the following collections.

I examined recipe books held by the following institutions:
   The British Library
   Chawton House Library
   The Folger Shakespeare Library
   The Harley Foundation
   National Library of Medicine (US)
   New York Academy of Medicine
   The Open Library
   Royal College of Physicians
   Society of Antiquaries, London
   Stanford University
   University of Aberdeen
   University of Glasgow
   University of Pennsylvania
   Warwickshire Record Office
   Wellcome Library

I examined non-recipe book materials at the following institutions:
   Bedfordshire Archives and Records Service
   Bodleian Library, University of Oxford
   Bristol Archives
   Chelsea Physic Garden
   Devon Record Office, Wolborough Feoffees and Widows' Charity
      Papers
   Hammersmith and Fulham Local Archives
   Lambeth Palace Library, including Fairhurst Papers, Miscellaneous
      Papers of John Selden and Sir Matthew Hale
   London Metropolitan Archives, including Sutton's Hospital/
      Charterhouse papers, Diocese of London Papers, St Paul's
      Cathedral, Dean and Chapter papers

Sheffield City Archives, including Duke of Norfolk's Estate, Arundel
    (Arundel Castle Records)
Society of Antiquaries, London
The British Library Western Manuscripts, including the Sloane
    Collection
The Royal College of Physicians
The Royal Society
University of Aberdeen Special Collections
University of Cambridge libraries
University of Glasgow Special Collections, including the Hunterian
    Collection
Wadham College, University of Oxford
Warwickshire County Record Office, including the Waller Family of
    Woodcote Papers, the Warwick Borough Council Papers, Warwick
    Craft Guilds and Mysteries Papers
Wellcome Library
Wiltshire and Swindon Record Office, Yerbury Almshouses Papers

CHAPTER BIBLIOGRAPHIES

Introduction

Secondary Sources

Linda Alcoff and Elizabeth Potter, "Introduction: When Feminisms Intersect Epistemology", in Linda Alcoff and Elizabeth Potter (eds), *Feminist Epistemologies* (Routledge, 1993), pp. 1–14.

Anne Barrett, "Where Are the Women? How Archives Can Reveal Hidden Women in Science", in Claire G. Jones, Alison E. Martin, and Alexis Wolf (eds), *The Palgrave Handbook of Women in Science since 1600* (Palgrave, 2022), pp. 129–47.

Claire G. Jones, Alison E. Martin, and Alexis Wolf, "Women in the History of Science: Frameworks, Themes, and Contested Perspectives", in Claire G. Jones, Alison E. Martin, and Alexis Wolf (eds), *The Palgrave Handbook of Women in Science since 1600* (Palgrave, 2022), pp. 3–24.

Primary Sources

John James Audubon, *The Birds of America* (London, 1827–1838).

Elizabeth Blackwell, *A Curious Herbal Containing Five Hundred Cuts of the Most Useful Plants Which Are Now Used in the Practice of Physick*, vols I and II (London, 1737/1739).

Maria Sibylla Merian, *Metamorphosis Insectorum Surinamensium* (Amsterdam, 1705).

PART ONE
1. Kitchen Physic Is the Best Physic

Secondary Sources

Edward Bever, "Witchcraft, female aggression, and power in the early modern community", *Journal of Social History* 35, n. 4 (2002), pp. 955–88.

Deborah Coltham, *"Ladies in the Laboratory"*: A Chronological List of Books by, or Relating to Women in Medicine and Science, Recent Acquisitions Five (Deborah Coltham Rare Books, 2009).

Allen G. Dubus, "Chemists, physicians, and changing perspectives on the scientific revolution", *Isis* 89, n. 1 (1998), pp. 61–81.

Peter Elmer, *Witchcraft, Witch-Hunting, and Politics in Early Modern England* (Oxford University Press, 2016).

Gerry Greenstone, "The history of bloodletting", *British Columbia Medical Journal* 52, n. 1 (2010), pp. 12–14.

Anna Keay, *The Restless Republic: Britain without a Crown* (HarperCollins, 2021).

Molly McClain, "The Duke of Beaufort's Tory Progress through Wales, 1684", *Cylchgrawn Hanes Cymru/Welsh History Review* 18, n. 4 (1997), pp. 593–620.

Steven Shapin and Simon Shaffer, *Leviathan and the Air Pump: Hobbes, Boyle, and the Experimental Life. Including a Translation of Thomas Hobbes, Dialogus Physicus de Natura Aeris, by Simon Shaffer* (Princeton University Press, 1985).

Patrick Wallis, "Plagues, morality and the place of medicine in early modern England", *English Historical Review* 121, n. 490 (2006), pp. 1–24.

A.S. Weber, "Women's early modern medical almanacs in historical context", *English Literary Renaissance* 33, n. 3 (2003), pp. 358–402.

Christopher J.M. Whitty, "British Books and Books Published in English Related to Medicine, 1475–1640: A Handlist of Identified Works", *The Medical World of Early Modern England, Wales and Ireland, 1500–1715*, Working Paper No. 3.

Primary Sources

Anon., *The Accomplish'd Ladies Rich Closet of Rarities* (London, 1687).

Francis Bacon, *Novum Organum (1620)*, Joseph Devey (ed.) (P. F. Collier, 1902).

Francis Bacon, *The Great Instauration in The Advancement of Learning (1605)*, Joseph Devey (ed.) (P. F. Collier, 1910).

[Eleazar] Duncan, *Wholesome Advice Against the Abuse of Hot Liquors* (London, 1706).

John Fothergill, *Lectures on the Materia Medica*, vol. II [only] (Wellcome Library/London, before 1754).

Robert Green, *A Quip Upon the Courtier* (London, 1592).

William Harvey, *Exercitatio Anatomica de Motu Cordis et Sanguinis in Animalibus (On the Motion of the Heart and Blood in Animals)* (London, 1628).

I.M., *A Proper New Booke of Cookery* (London, 1575).

William Lawson, *A New Orchard and Garden with The Country Housewifes Garden (1618)*, a facsimile edition with an introduction by Malcolm Thick (Prospect Books, 2003).

Mary Trye, *Medicatrix, Or The Woman-Physician* (London, 1675).

Andreas Vesalius, *De humani corporis fabrica libri septem* (Basel, 1543).

Christopher Wirtzung, *The General Practise of Physicke* (London, 1605).

2. The Countess of Kent's Recipe Book

Secondary Sources

A.D. Boney, *The Lost Gardens of Glasgow University* (Christopher Helm, 1988).

James Fitzmaurice, "Jane Barker and the tree of knowledge at Cambridge University", *Renaissance Forum* 3, n. 1 (1998), unpaginated.

Antonia Fraser, *The Wives of Henry VIII* (Alfred A. Knopf, 1993).

Patricia Higgins, "Grey, Elizabeth, Countess of Kent (1581–1651)", in Cathy Hartley (ed.), *A Historical Dictionary of British Women* (Routledge, 2003), p. 194.

Lynette Hunter, "Women and Domestic Medicine: Lady Experimenters, 1570–1620", in Lynette Hunter and Sarah Hutton (eds), *Women, Science and Medicine 1500–1700* (Sutton, 1997), pp. 89–107.

Lynette Hunter, "Women and Science in the Sixteenth and Seventeenth Centuries", in Judith P. Zinsser (ed.), *Men, Women, and the Birthing of Modern Science* (Northern Illinois University Press, 2005), pp. 123–40.

Peter Lely, *John Selden (1584–1654), c.* 1644. Oil on canvas, 28 × 24 inches (71.12 × 60.96 cm). Yale University Art Gallery. See also https://

artgallery.yale.edu/collections/objects/47608; accessed 1 Jan. 2021.

Molly McClain, "The Duke of Beaufort's Tory Progress through Wales, 1684", *Cylchgrawn Hanes Cymru/Welsh History Review* 4 (1997), pp. 593–620.

Jennifer Rabe, "Mediating between Art and Nature: The Countess of Arundel at Tart Hall", in Susanna Burghartz, Lucas Burkart, and Christine Gottler (eds), *Sites of Mediation: Connected Histories of Places, Processes, and Objects in Europe and Beyond, 1450–1650* (Brill, 2007), pp. 183–210.

Betty Travitsky and Anne Lake Prescott, *Seventeenth-Century English Recipe Books: Cooking, Physic and Chirurgery in the Works of W.M. and Queen Henrietta Maria, and of Mary Tillinghast* (Routledge, 2008).

Paul Van Somer, *Lady Elizabeth Grey, Countess of Kent. c.* 1619. Oil paint on wood, 1,143 × 819 mm. Tate Gallery.

Wendy Wall, "Literacy and the domestic arts", *Huntington Library Quarterly* 73, n. 3 (2010), pp. 383–412.

Primary Sources

Anon., *The Good Huswife's Handmaide for the Kitchin* (London, 1594).

Anon., *The Widowe's Treasure* (London, 1588).

Anon., *The Ladies Dictionary* (London, 1694).

Thomas Brugis, *The Marrow of Physicke* (London, 1648).

William Bullein, *Bulleins Bulwarke of Defence Against all Sicknesse, Soarenesse, and Woundes that Doe Dayly Assaulte Mankinde* (London, 1562).

Elizabeth Grey, Countess of Kent (attr.), *A Choice Manual of Rare and Select Secrets in Physick and Chryrurgery* (London, 1653).

Henrietta Maria of France (attr.), *The Queen's Closet Opened* (London, 1655).

Alethea Howard, Countess of Arundel and Lennox (attr.), *Natura Exenterata* (London, 1655).

L.M., *Prepositas His Practice* (London, 1588).

Queen Henrietta Maria (attr.), *The Queen's Closet Opened* (London, 1655).

Gervase Markham, *The English Hous-Wife, Containing The inward and outward Vertues which ought to be in a compleat Woman* (London, 1656).

Robert May, *The Accomplish'd Cook* (London, 1660).

Thomas Newton, *Approv'd Medicines and Cordial Receipts* (London, 1580).

Andrewe Plowden, "A Book of Surgerie and Phisick of Mistress Honorie Henslow", Manuscripts, 1601, http://WDAgo.com/s/791cf81b. Wiley Digital Archives: The Royal College of Physicians; accessed 1 Oct. 2021.

Jane Sharp, *The Midwives Book* (London, 1671).

John Smith, *England's Improvement Reviv'd: Digested into Six Books* (London, 1670).

Salvator Winter and Francisco Dickinson, *A Pretious Treasury* (London, 1649).

Hannah Woolley, *The Ladies Directory, in Choice Experiments & Curiosities of Preserving in Jellies, and Candying Both Fruits & Flowers* (London, 1662).

Owen Wood, *An Alphabetical Book of Physicall Secrets* (London, 1639).

3. Chicken Soup and Viper Wine

Secondary Sources

Rotimi Adigun, Hajira Basit, and John Murray, "Cell Liquefactive Necrosis", *StatPearls* (20 Aug. 2021).

Nick Bailey, *The Chelsea Physic Garden* (Chelsea Physic Garden, 2015).

E.W. Bligh, *Sir Kenelm Digby and His Venetia* (S. Low, Marston & Co., 1932).

Rose Bradley, *The English Housewife in the Seventeenth & Eighteenth Centuries* (E. Arnold, 1912).

J. Burnby, "Some early London physic gardens", *Pharmaceutical Historian* 24, n. 4 (1994), pp. 2–8.

Cambridge University Botanic Garden, University of Cambridge Museums and Botanic Garden, *History of the Garden, 1762–Present*. See also www.botanic.cam.ac.uk; accessed 21 Mar. 2020.

Julie Davis, "Botanizing at Badminton House: The Botanical Pursuits of Mary Somerset, First Duchess of Beaufort", in Donald L. Opitz, Staffan Bergwik, and Birgitte Van Tiggelen (eds), *Domesticity in the Making of Modern Science* (Palgrave Macmillan, 2016), pp. 19–40.

Margaret DeLacy, *The Germ of an Idea: Contagionism, Religion, and Society in Britain 1616–1730* (Palgrave Macmillan, 2016).

B.J. Dobbs, "Studies in the natural philosophy of Sir Kenelm Digby", parts I–IV, *Ambix* 20 (1973–1974), pp. 143–63.

Mordechai Feingold, *The Mathematicians' Apprenticeship: Science, Universities and Society in England, 1560-1640* (Cambridge University Press, 1984).

Michael Foster, "Digby, Sir Kenelm (1603–1665)", in *Oxford Dictionary of National Biography Online*, Oxford, 2009, https://doi.org/10.1093/ref:odnb/7629; accessed 21 Dec. 2021.

John F. Fulton, *Sir Kenelm Digby: Writer, Bibliophile and Protagonist of William Harvey* (Peter & Katherine Oliver, 1937).

Eric Griffin, "Copying 'the Anti-Spaniard': Post-Armada Hispanophobia and English Renaissance Drama", in Barbara Fuchs and Emily Weissbourd (eds), *Representing Imperial Rivalry in the Early Modern Mediterranean* (University of Toronto Press, 2015), pp. 191–216.

Lynette Hunter, "Women and Domestic Medicine: Lady Experimenters, 1570–1620", in Lynette Hunter and Sarah Hutton (eds), *Women, Science and Medicine 1500–1700* (Sutton, 1997), pp. 89–107.

Lynette Hunter, "Women and Science in the Sixteenth and Seventeenth Centuries", in Judith P. Zinsser (ed.) *Men, Women, and the Birthing of Modern Science* (Northern Illinois University Press, 2005), pp. 123–40.

Jardí Botànic de la Universitat de València, Botanic Gardens Conservation International, "History". See also https://www.jardibotanic.org/?apid=historia; accessed 21 Mar. 2020.

Molly McClain, *Beaufort: The Duke and his Duchess 1657–1715* (Yale University Press, 2001).

Sam A. Mellick, "Sir Kenelm Digby (1603–1665): diplomat, entrepreneur, privateer, duellist, scientist and philosopher", *ANZ Journal of Surgery* 81 (2011), pp. 911–14.

Wyndham Miles, "Sir Kenelm Digby, Alchemist, Scholar, Courtier, and

Man of Adventure", *Chymia* 2 (1949), pp. 119–28.

Sue Minter, *The Apothecaries' Garden: A History of the Chelsea Physic Garden* (Sutton, 2000).

Joe Moshenska, *A Stain in the Blood: The Remarkable Voyage of Sir Kenelm Digby* (Windmill, 2016).

John Parker, "The development of the Cambridge University botanic garden", *Curtis's Botanical Magazine* 23, n. 1 (2006), pp. 4–19.

Jennifer Rabe, "Mediating between Art and Nature: The Countess of Arundel at Tart Hall", in Susanna Burghartz, Lucas Burkart, and Christine Gottler (eds), *Sites of Mediation: Connected Histories of Places, Processes, and Objects in Europe and Beyond, 1450–1650* (Brill, 2007), pp. 183–210.

Royal College of Physicians of Edinburgh, "The Surgeons' Curriehill House – From Plants to Body-snatchers". See also www.rcpe.ac.uk/heritage/surgeons-curriehill-house-plants-body-snatchers; accessed 21 Mar. 2021.

Rachel Savage, "The herbal tradition and its influence on women's role in garden-making, 1600–1900", *Garden History* 46, n. 1 (2018), pp. 57–73.

Ann Shteir, *Cultivating Women, Cultivating Science: Flora's Daughters and Botany in England* (The Johns Hopkins University Press, 1996).

Anna Simmons, "Medicines, monopolies and mortars: the chemical laboratory and pharmaceutical trade at the Society of Apothecaries in the eighteenth century", *Ambix* 53, n. 3 (2006), pp. 221–36.

Helen Smith, "Eggs, Cheese, and (Francis) Bacon", in Barbara Fuchs and Emily Weissbourd (eds), *Representing Imperial Rivalry in the Early Modern Mediterranean* (University of Toronto Press, 2015), pp. 140–66.

Richard Sugg, *Mummies, Cannibals, and Vampires: The History of Corpse Medicine from the Renaissance to the Victorians* (Routledge, 2011).

Elaine Tierney, "'Dirty rotten sheds': exploring the ephemeral city in early modern London", *Eighteenth-Century Studies* 50, n. 2 (2017), pp. 231–52.

Katherine Tyrrell, "Botanic and Physic Gardens of the Past in London".

Botanical Art & Artists. See also www.botanicalartandartists.com; accessed 20 Mar. 2021.

University of Arizona Health Sciences, "Brain Liquefaction After Stroke is Toxic to Surviving Brain", *ScienceDaily* (20 Feb. 2018).

University of Glasgow, "A Significant Medical History: 18th Century". See also https://www.gla.ac.uk/schools/medicine/mus/ourfacilities/history/; accessed 19 July 2019.

Simon Werrett, *Thrifty Science: Making the Most of Materials in the History of Experiment* (University of Chicago Press, 2019).

Margaret Willes, *The Making of the English Gardener: Plants, Books and Inspiration, 1560–1660* (Yale University Press, 2011).

Margaret Willes, *The Domestic Herbal: Plants for the Home in the Seventeenth Century* (Bodleian Library Publishing, 2020).

Andrea Wulf, *The Brother Gardeners: Botany, Empire and the Birth of an Obsession* (Alfred A. Knopf, 2008).

Primary Sources

By or attributed to Kenelm Digby in chronological order:

A Discourse Concerning the Vegetation of Plants Spoken by Sir Kenelm Digby at Gresham College, on the 23 of January 1660 (London, 1661).

Choice and Experimented Receipts in Physic and Chirurgery, as also Cordial and Distilled Waters and Spirits, Perfumes, and other Curiosities, Collected by the Honourable and truly Learned Sir Kenelm Digby, Kt (London, 1668).

The Closet of Sir Kenelm Digby Kt Opened: Whereby is Discovered Several ways for making of Metheglin, Sider, Cherry-Wine, etc. Together with Excellent Directions for Cookery: As also for Preserving, Conserving, Candying, etc (London, 1669).

Two Treatises, By the Honourable and truly Learned Sir Kenelm Digby Knight. The one, Of Choice and Experimented Receipts in Physic and Chirurgery; as also Cordial and Distilled Waters and Spirits, Perfumes, and other Curiosities. The other, Of Cookery, With several ways for Making of Metheglin, Sider, Cherry-Wine, etc. Together with Excellent Directions for Preserving, Conserving, Candying, etc (London, 1669).

Choice and Experimented Receipts in Physic and Chirurgery... (London, 1675).

A Choice Collections of Rare Secrets and Experiments in Philosophy, as also Rare and unheard-of Medicines, Menstruums, and Alkahests; with the True Secret of Volatilizing the fixt Salt of Tartar. Collected And Experimented by the Honourable and truly Learned Sir Kenelm Digby, Kt (London, 1682).

Chymical Secrets and Rare Experiments in Philosophy (London, 1683).

By others, in alphabetical order:

Anon., *The Ladies Cabinet Opened, Wherein is Found Hidden Several Experiments in Preserving and Conserving, Physick, and Surgery, Cookery, and Huswifery* (London, 1639).

John Aubrey, in Kate Bennett (ed.) *John Aubrey: Brief Lives with An Apparatus for the Lives of our English Mathematical Writers*, vols I (text) and II (commentary) (Oxford, 2016).

Philip Bellon, *The Potable Balsome of Life* (London, 1675).

Elizabeth Blackwell, *A Curious Herbal Containing Five Hundred Cuts of the Most Useful Plants Which Are Now Used in the Practice of Physick*, vols I and II (London, 1737/1739).

Robert Boyle, *Work diaries* (Royal College of Physicians/London).

Henry Bracken, *The Midwife's Companion* (London, 1737).

Richard Bradley, *New Improvements of Planting and Gardening* (London, 1716).

Charles Carter, *The Compleat City and Country Cook: or, Accomplish'd Housewife* (London, 1732).

Nicholas de Bonnefons, *Le Jardinier François* (Paris, 1651).

John Donne, "Sermon X. Preached upon Candlemas Day", in *LXXX sermons preached by that learned and reverend divine, John Donne, Dr in Divinity, late Deane of the cathedrall church of S. Pauls London* (London, 1640).

William Dover, *Useful Miscellanies* (London, 1739).

Charles Evelyn, *The Art of Gardening, Improv'd* (London, 1717).

John Evelyn (trans.), *The French Gardiner* (London, 1658).

F.B., *The Office of the Good House-wife, With Necessary Directions for the Ordering of her Family and Dairy* (London, 1672).

Thomas Fuller, *A Pisgah-sight of Palestine* (London, 1650).

Sarah Harrison, *The House-keeper's Pocket-Book And Compleat Family Cook* (London, 1739).

Alethea Howard, Countess of Arundel and Lennox (attr.), *Natura Extenterata* (London, 1655).

John Jones, *Adrasta: Or, The Womans Spleene, And Loves Conquest* (London, 1635).

Batty Langley, *New Principles of Gardening*, 2nd edn (London, 1739).

Christopher Langton, *An Introduction into Phisycke, with an Universal Dyet* (London, 1545).

William Lawson, *A New Orchard and Garden with The Country Housewifes Garden (1618)*, a facsimile edition with an introduction by Malcolm Thick (Prospect Books, 2003).

Matthew Mackaile, *Macis macerata: or, a Short Treatise, Concerning the use of mace, in meat, or drink, and medicine* (Aberdeen, 1677).

Robert May, *The Accomplish'd Cook* (London, 1660).

John Middleton, *Five Hundred New Receipts in Cookery, Confectionary, Pastry, Preserving, Conserving, Pickling, and The Several Branches of These Arts Necessary To Be Known By All Good Housewives* (London, 1734).

James Millerd, *An Exact Delineation of the Famous Citty of Bristoll and Suburbs* (Bristol, 1673).

Thomas Newton, *Approv'd Medicines and Cordial Receipts* (London, 1580).

Thomas Newton, *The Olde Mans Dietarie* (London, 1586).

John Nott, *The Cook's and Confectioner's Dictionary* (London, 1723).

T.P. (Hannah Woolley attr.), *The Accomplish't-Ladys Delight in Preserving, Physick, Beautifying, and Cookery* (London, 1675).

Dorothy Partridge, *Woman's Almanack for the Year 1694* (London, 1694).

John Partridge, *The Treasurie of Commodious Conceits, & Hidden Secrets* (London, 1573).

John Partridge, *The Widowes Treasure, Plentifully Furnished with Secretes in Phisicke* (London, 1586).

Richard Poulteney, *Historical and Biographical Sketches of the Progress of Botany in England* (London, 1790).

Francis Quarles, "Meditation 10", in *Divine Poems* (London, 1633).

Alexander Read, *The Chirurgicall Lectures of Tumors and Ulcers* (London, 1635).

William Salmon, *Iatrica: Seu Praxis Medendi. The Practice of Curing* (London, 1681).

John Smith, *England's Improvement Reviv'd* (London, 1670).

Henry Stevenson, *The Young Gardener's Director* (London, 1716).

Stephen Switzer, *The Practical Kitchen Gardiner* (London, 1727).

Thomas Tryon, *The Good Housewife made a Doctor* (London, 1692).

Thomas Tryon, *A Pocket-Companion* (London, 1693).

Tobias Whittaker, *The Tree of Humane Life, or, The Bloud of the Grape* (London, 1634).

Thomas Willis, *Pharmaceutice rationalis: or, An Exercitation of the Operations of Medicines in Humane* (London, 1678).

4. Proscriptions, Prescriptions, and Poetry

Secondary Sources

Katherine Allen, "Recipe collections and the realities of fashionable diseases in eighteenth-century elite domestic medicine", *Literature and Medicine* 35, n. 2 (2017), pp. 334–54.

Jonathan Barry, "The 'compleat physician' and experimentation in medicines: Everard Maynwaring (*c.* 1629–1713) and the Restoration Debate on Medical Practice in London", *Medical History* 62, n. 2 (2018), pp. 155–76.

Sarah Birt, "Women, guilds and the tailoring trades: the occupational training of Merchant Taylors' company apprentices in early modern London", *London Journal* (2020), pp. 1–19.

Isabelle Clairhout, "Erring from good huswifry? The author as witness in Margaret Cavendish and Mary Trye", *Renaissance and Reformation/ Renaissance et Réforme* 37, n. 2 (2014), pp. 81–114.

Harold J. Cook, "The Society of Chemical Physicians, the New Philosophy, and the Restoration Court", *Bulletin of the History of Medicine* 61, n. 1 (1987), pp. 61–77.

James Fitzmaurice, "Jane Barker and the tree of knowledge at Cambridge University", *Renaissance Forum* 3, n. 1 (1998), unpaginated.

James Fitzmaurice, "Daring and innocence in the poetry of Elizabeth Rochester and Jane Barker", *In-Between: Essays & Studies in Literary Criticism* 11, n. 1 (2002), pp. 25–43.

Elizabeth Lane Furdell, *The Royal Doctors, 1485–1714: Medical Personnel at the Tudor and Stuart Courts* (Boydell & Brewer, 2001).

Karen Bloom Gevirtz, "Philosophy and/in Verse: Jane Barker's 'Farewell to Poetry' and the Anatomy of Emotion", in Robin Runia (ed.), *The Future of Feminist Eighteenth-Century Scholarship: Beyond Recovery* (Routledge, 2017), pp. 53–70.

Kathryn King, *Jane Barker, Exile: A Literary Career 1675–1725* (Clarendon Press, 2000).

Kathryn R. King, "Of needles and pens and women's work", *Tulsa Studies in Women's Literature* 14, n. 1 (1995), pp. 77–93.

Stanton J. Linden, "Mrs Mary Trye, medicatrix: chemistry and controversy in restoration England", *Women's Writing* 1, n. 3 (1994), pp. 341–53.

Niall MacKenzie, "Jane Barker, Louise Hollandine of the Palatinate and 'Solomons [sic] Wise Daughter'", *Review of English Studies*, New Series 58, n. 233 (2007), pp. 64–72.

Marjorie H. Nicolson, "Ward's 'Pill and Drop' and men of letters", *Journal of the History of Ideas* 29, n. 2 (1968), pp. 177–96.

Sara Read, "'My method and medicines': Mary Trye, Chemical Physician", *Early Modern Women* 11, n. 1 (2016), pp. 137–48.

Philip Walsingham Sergeant, *My Lady Castlemaine: Being a Life of Barbara Villiers, Countess of Castlemaine, Afterwards Duchess of Cleveland* (Hutchinson, 1912).

Deborah Simonton, "Toleration, Liberty and Privileges: Gender and Commerce in Eighteenth-Century European Towns", in Deborah Simonton (ed.), *The Routledge History Handbook of Gender and the Urban Experience* (Routledge, 2017), pp. 33–46.

Richard Sugg, *Mummies, Cannibals, and Vampires: The History of Corpse Medicine from the Renaissance to the Victorians* (Routledge, 2011).

Angela Vanhaelen and Bronwen Wilson (eds), *Making Worlds: Global*

Invention in the Early Modern Period (University of Toronto Press, 2022), pp. 3–21.

Patrick Wallis, "Plagues, morality and the place of medicine in early modern England", *English Historical Review* 121, n. 490 (2006), pp. 1–24.

Primary Sources

Anon. *The Present State of Physick & Surgery in London* (London, 1701).

Jane Barker, *The Galesia Trilogy and Selected Manuscript Poems of Jane Barker*, Carol Shiner Wilson (ed.) (Oxford University Press, 1997).

Jane Barker, *Poems on Several Occasions (1688)*. Magdalen College Library, University of Oxford. *The Perdita Project*, University of Warwick.

Daniel Bellamy, "The Merry Swain" in *Miscellanies in Prose and Verse, Consisting of Dramatick Pieces, Poems, Humorous Tales, Fables, &c. by Daniel Bellamy*. vol 2 (London, 1739).

Robert Boyle, *A Free Inquiry into the Vulgarly Receiv'd Notion of Nature* (London, 1681).

James Bramston, *The Man of Taste. Occasion'd by an Epistle of Mr. Pope's on That Subject* (London, 1733).

Thomas Brugis, *The Marrow of Physicke* (London, 1648).

John Colbatch, *The Generous Physician, or Medicine Made Easy* (London, 1733).

Roger Dixon, *Consultum Sanitatis, A Directory to Health* (London, 1663).

Michael Drayton, *Englands Heroical Epistles* (London, 1689).

Esther Dudley, "Certificate of admission of Esther Dudley to the freedom of the City of London", (1741). Waller Family of Woodcote Papers (Warwickshire Record Office).

Thomas Dyche and William Patton, *A New General English Dictionary* (London, 1740).

John Fothergill, *Lectures on the Materia Medica*, vol II [only] (Wellcome Library, London).

Thomas Fuller, *A Pisgah-sight of Palestine* (London, 1650).

Thomas Fuller, *Pharmacopoeia extemporanea*, 2nd edn (London, 1714).

James Gillray, *The Gout*. 14 May 1799. Hand-coloured etching and aquatint, 260 mm × 355 mm (Metropolitan Museum of Art).

Alexander Gordon, *Lecture notes*. University of Aberdeen Special Collections (University of Aberdeen).

Hermannus Vander Heyden, *Speedy Help for Rich and Poor* (London, 1653).

James Howell, *Paroimiographia Proverbs* (London, 1659).

Nicolas Le Fèvre, *A Discourse Upon Sir Walter Rawleigh's Great Cordial* (London, 1778).

Richard Lower, *Tractatus de Corde: Item de Motu & Colore Sanguinis et Chyli in cum Transit* (London, 1669).

John Marten, *The Attila of the Gout*, 2nd edn (London, 1713).

E. Maynewaring, *A Treatise of Consumptions* (London, 1668).

A Merchant in London, *The Present State of Physick & Surgery in London* (London, 1701).

Roger Pitt, *The Craft and Frauds of Physick Expos'd*, 2nd edn (London, 1703).

John Quincy, *Pharmacopoeia Officinalis & Extemporanea. Or, A Complete English Dispensatory*, 5th edn (London, 1724).

John Quincy, *Lexicon Physico-Medicum: Or, A New Medicinal Dictionary* (London, 1726).

John Ray, *A Collection of English Proverbs* (Cambridge, 1670).

Royal College of Physicians of London, *Pharmacopoeia Londinensis* (London, 1702, 1721, 1724).

William Salmon, *The Family-Dictionary; Or, Houshold [sic] Companion*, 2nd edn (London, 1696).

Sophia, A Person of Quality, *Woman's Superior Excellence Over Man* (London, 1740).

Mary Trye, *Medicatrix, Or The Woman-Physician* (London, 1675).

Hermannus Vander Heyden, *Speedy Help for Rich and Poor* (London, 1653).

Benjamin Welles, *A Treatise of the Gout* (London, 1669).

Thomas Willis, *Cerebri Anatome, Cui Accessit Nervorum Descripto et Usus Studio Thomæ Willis* (London, 1664).

Thomas Willis, *Pathologiæ Cerebri, et Nervosi Generis Specimen in quo Agitur de Morbis Convulsivis, et de Scorbuto* (Oxford, 1667).

5. "Was Once a Science, Now's a Trade"

Secondary Sources

Stuart Anderson, "'A proper person to officiate': apothecaries at Westminster Hospital, London—1716 to 1826", *Pharmacy in History* 49, n. 1 (2007), pp. 3–14.

Attorney General's Office, "Attorney General Bonta Sues Nation's Largest Insulin Makers, Pharmacy Benefit Managers for Illegal Practices, Overcharging Patients", 23 Jan. 2023. See also https://oag.ca.gov/news/press-releases/attorney-general-bonta-sues-nations-largest-insulin-makers-pharmacy-benefit; accessed 4 Apr. 2023.

C. C. Booth, "Sir Samuel Garth, F.R.S.: The Dispensary Poet", *Notes and Records of the Royal Society* 40 (1985–6), pp. 125–45.

Jeremy Boulton, "Going on the Parish: The Parish Pension and its Meaning in the London Suburbs, 1640–1724", in Tim Hitchcock, Peter Searle, and Pamela Sharpe (eds), *Chronicling Poverty* (Macmillan, 1997), pp. 19–46.

Scott Breuninger, "A panacea for the nation: Berkeley's Tar-water and Irish domestic development", *Études irlandaises* 34, n. 2 (2009), pp. 29–41.

Richard I. Cook, *Sir Samuel Garth* (Twayne Publishers, 1980).

Richard Coulton, "'What he hath Gather'd Together Shall not be Lost': Remembering James Petiver", *Notes and Records* 74 (2020), pp. 189–211.

Frank H. Ellis, "The background of the London Dispensary", *Journal of the History of Medicine and Allied Sciences* 20, n. 3 (1965), pp. 197–212.

Karen Bloom Gevirtz, *Life after Death: Widows and the English Novel, Defoe to Austen* (University of Delaware Press, 2005).

David Harley, "'Bred up in the study of that faculty': licensed physicians in north-west England, 1660–1760", *Medical History* 38 (1994), pp. 398–420.

E.E. Harrison, *The History of the Charterhouse and its Buildings.*

Reprinted from the Transactions of the Ancient Monuments Society (1990).

William Hartston, "Medical dispensaries in eighteenth-century London" (abridged), *Proceedings of the Royal Society of Medicine* 56 (1963), pp. 753–58.

Arzu Babayigit Hocaoglu *et al.*, "Glycyrrhizin and long-term histopathologic changes in a murine model of asthma", *Current Therapeutic Research* 72, n. 6 (2011), pp. 250–61.

Muzaffar Iqbal, *The Making of Islamic Science* (The Other Press, 2009).

[Kennedy] Newsroom, "Kennedy, Warnock introduce bipartisan bill to cap insulin prices, lower cost of diabetic care", Office of Senator Kennedy, Senator for Louisiana, 23 Mar. 2023.

William Kerwin, "Where Have You Gone, Margaret Kennix? Seeking the Tradition of Healing Women in English Renaissance Drama", in Liliam R. Furst (ed.), *Women Healers and Physicians: Climbing a Long Hill* (University Press of Kentucky, 1997), pp. 93–113.

Yi Kuang *et al.*, "Antitussive and expectorant activities of licorice and its major compounds", *Organic & Medicinal Chemistry* 26, n. 1 (2018), pp. 278–84.

Joan Lane, "Provincial medical apprentices and masters in early modern England", *Eighteenth-Century Life* 12, n. 3 (1988), pp. 14–27.

Ephraim Philip Lansky *et al.*, "Ficus spp. (fig): Ethnobotany and potential as anticancer and anti-inflammatory agents", *Journal of Ethnopharmacology* 119, n. 2 (2008), pp. 195–213.

Zhongyuan Li *et al.*, "A comprehensive review on phytochemistry, bioactivities, toxicity studies, and clinical studies on Ficus carica Linn. Leaves", *Biomedicine & Pharmacotherapy* 137 (2021) 111393.

Stephen Porter, *The London Charterhouse* (Amberley Publishing, 2009).

Harriet Richardson, "Aberdeen", *Historic Hospitals: An Architectural Gazetteer*, 4 Feb. 2018. See also https://historic-hospitals.com/gazetteer/aberdeen/; accessed 17 July 2019.

Harriet Richardson, "Bristol Royal Infirmary", *Historic Hospitals: An Architectural Gazetteer*, 4 Feb. 2018. See also https://historic-hospitals.com/2018/02/04/bristol-royal-infirmary/; accessed 17 July 2019.

Albert Rosenberg, "The London Dispensary for the sick-poor", *Journal of the History of Medicine and Allied Sciences* 14, n. 1 (1959), pp. 41–56.

Cathy Ross, "'Men of Honour and Power': The Charterhouse's Restoration Governors", *The Great Chamber at the Charterhouse*. See also www.thecharterhouse.org; accessed 23 Feb. 2023.

Royal College of Physicians of Edinburgh, "The College Dispensary", Heritage. See also www.rcpe.ac.uk/heritage/college-history/college-dispensary; accessed 28 May 2023.

Royal College of Physicians of Edinburgh, "The Dispensary Movement", Heritage. See also www.rcpe.ac.uk/heritage/eighteenth-century-dispensary-movement; accessed 28 May 2023.

Bram Sable-Smith, "Insulin's Steep Price Leads To Deadly Rationing", *Kaiser Health News* (7 Sept. 2018).

Bisma A. Sayed *et al.*, "Insulin Affordability and the Inflation Reduction Act: Medicare Beneficiary Savings by State and Demographics" (US Department of Health and Human Services, 24 Jan. 2023).

John F. Sena, "Samuel Garth's The Dispensary", *Texas Studies in Literature and Language* 15, n. 4 (1974), pp. 639–48.

Kevin Siena, "Contagion, Exclusion, and the Unique Medical World of the Eighteenth-Century Workhouse: London Infirmaries in Their Widest Relief", in Jonathan Reinarz and Leonard Schwartz (eds), *Medicine and the Workhouse* (Boydell & Brewer, 2013), pp. 19–39.

Anna Simmons, "Medicines, monopolies and mortars: the chemical laboratory and pharmaceutical trade at the Society of Apothecaries in the eighteenth century", *Ambix* 53, n. 3 (2006), pp. 221–36.

Glenn Sonnedecker, "The Apothecary in a Scottish Infirmary", *Pharmacy in History* 41, n. 3 (1999), pp. 119–20.

Melissa Suran, "All 3 major insulin manufacturers are cutting their prices—Here's what the news means for patients with diabetes", *JAMA Medical News* 329, n. 16 (2023), pp. 1337–39.

G.J. Toomer, *Eastern Wisedome and Learning: The Study of Arabic in Seventeenth-Century England* (Clarendon Press, 2007).

[Warnock] Newsroom, "Senators Reverend Warnock, Kennedy Introduce Bipartisan Legislation to Cap Insulin Costs at $35 a Month for Everyone", Office of Reverend Raphael Warnock, US Senator for

Georgia, 23 Mar. 2023.

J.W. Willcock, *The Laws Relating to the Medical Profession* (J. and W.T. Clarke, 1830).

Matthew Yeo, *The Acquisition of Books in Chetham's Library, 1655–1700* (Brill, 2010).

Primary Sources

Anon., *The Present State of Physick and Surgery in London* (London, 1701).

Anon., *Siris in the Shades: A Dialogue Concerning Tar Water* (London, 1744).

Anon., "The Wandering Beauty", in Janet Todd (ed.), *The Works of Aphra Behn*, vol. 3 "The Fair Jilt and Other Short Stories" (London, 1995), pp. 1–48.

Thomas Brown, *Physick Lies A-Bleeding, or the Apothecary turned Doctor* (London, 1697).

Susannah Centlivre, *The Basset Table*, Jane Milling (ed.) (Broadview Press, 2009).

Hugh Chamberlen, *A Proposal For the better Securing of Health* (London, 1689).

Chymist in the City, *Bellum medicinale, or the present state of doctors and apothecaries in London* (London, 1701).

Anne Clifford, from "Diary, 1616–1617", in Elspeth Graham, Hilary Hinds, Elaine Hobby, and Helen Wilcox (eds), *Her Own Life: Autobiographical Writings by Seventeenth-Century Englishwomen* (London, 1989), pp. 35–53.

Samuel Garth, *The Dispensary. A Poem* (London, 1699).

Thomas Gouge, *The Surest & Safest Way of Thriving* (London, 1673).

Mary Huntington, "Mary Huntington [née Powell] to John Locke: Thursday, 15 January 1699", in Robert McNamee *et al.* (eds), *Electronic Enlightenment Scholarly Edition of Correspondence* (Oxford, 2018).

John Locke, "Second Treatise on Government", in Peter Laslett (ed.), *John Locke: Two Treatises on Government* (Cambridge, 1988), pp. 265–428.

Richard Mead (attr.), *The Triumvirate: Or, The Battel Among Physicians,*

2nd edn (London, 1719).

William Salmon, *A Rebuke to the Authors of the Blew-Book, Call'd The State of Physick in London* (London, 1698).

John Shirley, *The Accomplished Ladies Rich Closet*, 2nd edn (London, 1687).

Henry Stonecastle, *The Universal Spectator, and Weekly Journal: By Henry Stonecastle, of Northumberland, Esq*, n. 332 (20 July 1734).

6. The Laboratory on Cheesewell Street

Secondary Sources

The Aspirin Foundation, "The story of Aspirin – a versatile medicine with a long history". See also www.aspirin-foundation.com/history/the-aspirin-story/; accessed 9 June 2023.

Jonathan Berry, "John Houghton and medical practice in London *c.* 1700", *Bulletin of the History of Medicine* 92, n. 4 (2018), pp. 575–603.

Helen Brock, "James Douglas of the Pouch", *Medical History* 18, n. 2 (1974), pp. 162–72.

Sajed Chowdhury, "Introducing women's alchemical practices", *Early Modern Women: An Interdisciplinary Journal* 15, n. 2 (2021), pp. 89–92.

Deborah Coltham, *"Ladies in the Laboratory": A chronological list of books by, or relating to women in medicine and science*, Recent Acquisitions Five (Deborah Coltham Rare Books, 2009).

Harold J. Cook, "The Society of Chemical Physicians, the New Philosophy, and the Restoration Court", *Bulletin of the History of Medicine* 61, n. 1 (1987), pp. 61–77.

Michelle DiMeo, *Lady Ranelagh: The Incomparable Life of Robert Boyle's Sister* (University of Chicago Press, 2021).

Mordechai Feingold, *The Newtonian Moment: Isaac Newton and the Making of Modern Culture* (New York Public Library and Oxford University Press, 2004).

R.T. Gunther, *Early Science in Oxford*, vol I ,"Chemistry, Mathematics, Physics, and Surveying" (Oxford Historical Society, 1923).

Lynette Hunter, "Sisters of the Royal Society: The Circle of Katherine Jones, Lady Ranelagh", in Lynette Hunter and Sarah Hutton (eds), *Women, Science and Medicine 1500-1700: Mothers* (Sutton, 1997), pp. 178–97.

Lynette Hunter, "Women and Domestic Medicine: Lady Experimenters, 1570–1620", in Lynette Hunter and Sarah Hutton (eds), *Women, Science and Medicine 1500–1700* (Sutton, 1997), pp. 89–107.

Lynette Hunter, "Women and Science in the Sixteenth and Seventeenth Centuries", in Judith P. Zinsser (ed.), *Men, Women, and the Birthing of Modern Science* (Northern Illinois University Press, 2005), pp. 123–40.

Laura Miller, "Masculinity, Space, and Late Seventeenth-Century Alchemical Practices", in Mona Narain and Karen Gevirtz (eds), *Gender and Space in British Literature, 1660–1820* (Routledge, 2014), pp. 165–78.

Jonathan Miner and Adam Hoffhines, "The discovery of aspirin's antithrombotic effects", *Texas Heart Institute Journal* 34 (2007), pp. 179–86.

Mt. Sinai Hospital, "Oxalic Acid Poisoning". See also www.mountsinai.org/health-library/poison/oxalic-acid-poisoning; accessed 29 June 2023.

New Jersey Department of Health, "*Hazardous Substance Fact Sheet: Oxalic Acid*". See also www.nj.gov/health/eoh/rtkweb/documents/fs/1445.pdf; accessed 29 June 2023.

"Nicholas Lémery (1645–1716)", *Nature* 156 (1945), p. 598.

Jennifer M. Rampling, *The Experimental Fire: Inventing English Alchemy, 1300–1700* (University of Chicago Press, 2021).

Hilary Rose, *Foreword*, in Lynette Hunter and Sarah Hutton (eds), *Women, Science and Medicine 1500–1700* (Sutton, 1997), pp. xi–xx.

The Royal Collection Trust, "Charles II and the Royal Observatory Greenwich", Charles II: Art & Power. See also www.rct.uk/collection/themes/exhibitions/charles-ii-art-power/the-queens-gallery-buckingham-palace/charles-ii-and-the-royal-observatory-greenwich; accessed 5 June 2023.

K. Bryn Thomas, "James Douglas of the Pouch, 1675–1742", *British Medical Journal* (1960), pp. 1649–50.

Courtney Weiss Smith, *Empiricist Devotions: Science, Religion, and Poetry in Early Eighteenth-Century England* (University of Virginia Press, 2016).

Primary Sources

Aphra Behn, "The Emperor of the Moon: A Farce", in *The Cambridge Edition of the Works of Aphra Behn*, vol IV, plays 1682–1696, edited by Rachel Adcock *et al.* (Cambridge University Press, 2021), pp. 309–530.

Noah Biggs, *The New Dispensatory* (London, 1651).

Elizabeth Blackwell, *A Curious Herbal Containing Five Hundred Cuts of the Most Useful Plants Which Are Now Used in the Practice of Physick*, vols I and II (London, 1737/1739).

Brockman Papers (British Library/London).

Susanna Centlivre, *The Basset Table*, Jane Milling (ed.) (Broadview Press, 2009).

Daniel Coxe, "'Of the effect of tobacco-oyle" (Royal Society/London).

Henry Curzon, *The Universal Library: Or, Compleat Summary of Science. Containing above sixty select treatises* (London, 1712).

James Douglas, *Letters*. Sloane Manuscripts (British Library/London).

James Douglas, "The natural history and description of the phoenicopterus or flamingo; with two views of the head, and three of the tongue, of that beautiful and uncommon bird", *Philosophical Transactions* 29, n. 350 (31 Dec. 1716), pp. 523–41.

Sarah Draper, *Recipe book* (Wellcome Library/London).

Richard Fletcher, *A Vindication of Chymistry, and Chymical Medicines* (London, 1676).

John Friend, *Chymical lectures* (London, 1712).

Hunterian Collection. University of Glasgow Special Collections (University of Glasgow Library/Glasgow).

Dr [John] Huxham [?], *English Medical Notebook* (Wellcome Library/London).

George Martine, *Essays medical and philosophical by George Martine, M.D.* (London, 1740).

Dorothea Rousby, *A Cookery Book with Index* [1694] (Stanford

University Library).

Thomas Shadwell, *The Virtuoso*, Marjorie Hope Nicolson and David Stuart Rhodes (eds) (University of Nebraska Press, 1966).

Peter Shaw, *Philosophical Principles* (London, 1730).

Peter Shaw, *Chemical Lectures* (London, 1734).

George Starkey, *A Necessary and Full Apology for Chymical Medicaments* (London, 1656).

Unknown authors, *Medical Accounts* (Wellcome Library/London).

Waller Family of Woodcote Papers (Warwickshire County Records Office).

George Wilson, *A Compleat Course of Chemistry, Containing near Three Hundred Operations; Several of which have not been seen before* (London, 1703).

7. The Doctoress's Cure for the Stone

Secondary Sources

Anon., "Nicolas Lémery (1645–1715)", *Nature* 156 (1945), p. 598.

Juanita G.L. Burnby, "The apothecary as progenitor", *Medical History* 27, n. S3 (1983), pp. 24–61.

Stephen Clucas, "Joanna Stephens's Medicine and the Experimental Philosophy", in Judith P. Zinsser (ed.), *Men, Women, and the Birthing of Modern Science* (Northern Illinois University Press, 2005), pp. 141–58.

John D. Comrie, "English medicine in the eighteenth century", *Proceedings of the Royal Society of Medicine* XXVII, pp. 37–44.

Penelope J. Corfield, "From poison peddlers to civic worthies: the reputation of the apothecaries in Georgian England", *Social History of Medicine* 22, n. 1 (2009), pp. 1–21.

B.J. Dobbs, "Studies in the natural philosophy of Sir Kenelm Digby. Part III. Digby's experimental alchemy–the book of secrets", *Ambix* 21, n. 1 (1974), pp. 1–28.

Christopher Duffin, "Joseph Clutton, *c.* 1695-1743: A Georgian apothecary", *Pharmaceutical Historian* 48, n. 4 (2018), pp. 85–98.

Sandy Feinstein, "Experience, Authority, and the Alchemy of Language:

Margaret Cavendish and Marie Meurdrac Respond to the Art", *Early Modern Women: An Interdisciplinary Journal* 15, n. 2 (2021), pp. 133–42.

Elizabeth Lane Furdell, *The Royal Doctors, 1485–1714: Medical Personnel at the Tudor and Stuart Courts* (Boydell & Brewer, 2001).

David Harley, "'Bred Up in the Study of that Faculty': Licensed Physicians in North-West England, 1660–1760", *Medical History* 38 (1994), pp. 398–420.

J. Cordy Jeaffreyson, "Doctors out of Practice", *The Leisure Hour* (1884), pp. 346–50.

William Kerwin, "Where Have You Gone, Margaret Kennix? Seeking the Tradition of Healing Women in English Renaissance Drama", in Liliam R. Furst (ed.), *Women Healers and Physicians: Climbing a Long Hill* (University Press of Kentucky, 1997), pp. 99–113.

Edward L. Keys, "The Joanna Stephens Medicines for the Stone: A Faith that Failed", *The Bulletin* (1942), pp. 835–40.

Marjorie H. Nicolson, "Ward's 'Pill and Drop' and men of letters", *Journal of the History of Ideas* 29, n. 2 (1968), pp. 177–96.

Margaret Pelling and Frances White, *Physicians and Irregular Medical Practitioners in London Database* (London, 2004).

Eric Riches, "The history of lithotomy and lithotrity", *Annals of the Royal College of Surgeons of England* 43, n. 4 (1968), pp. 185–99.

Eric Riches, "Samuel Pepys and his stones", *Annals of the Royal College of Surgeons of England* 59 (1977), pp. 11–16.

Romney R. Sedgwick, "Hanbury Williams, Charles (1708–59), of Coldbrook, Mon.", in R. Sedgwick (ed.), *The History of Parliament: the House of Commons 1715-1754*, vol I (Boydell & Brewer, 1970). See https://www.historyofparliamentonline.org/volume/1715-1754/member/hanbury-williams-charles-1708-59; accessed 29 June 2023.

A History of the County of Lancaster: Volume 3, William Farrer and J Brownbill (eds.) (London, 1907).

Arthur J. Viseltear, "Joanna Stephens and the eighteenth century lithontriptics; misplaced chapter in the history of therapeutics", *Journal of the Bulleting of the History of Medicine* 42, n. 3 (1968), pp. 199–220.

Patrick Wallis, "Consumption, retailing, and medicine in early modern England", *Economic History Review* 61, n. 1 (2008), 26–53.

Wayne Wild, *Medicine-by-post: The Changing Voice of Illness in Eighteenth-century British Consultation Letters and Literature* (Brill, 2006).

D. Williams, "Williams, Sir Charles Hanbury (1708–1759), satirical writer and diplomatist", *Dictionary of Welsh Biography Online.*

A. Dickson Wright, "Quacks through the ages", *Journal of the Royal Society of Arts* 105, n. 4995 (1957), pp. 161–78.

Primary Sources
Newspapers
Athenian Gazette
Common Sense or The Englishman's Journal
Country Journal or The Craftsman
Daily Courant
Fog's Weekly Journal
Gentleman's Magazine
Grub Street Journal
London Daily Post
London Evening Post
Post Man and the Historical Account
Universal Spectator
Weekly Oracle or Universal Library

Other sources
"An act for providing a Reward to Joanna Stephens upon a proper Discovery to be made by her for the Use of the Publick, of the Medicines Prepared by her for the cure of the Stone" (Royal College of Physicians/London, 1741).

Anon., *The Country Physician* (Edinburgh, 1701).

Anon., *Siris in the Shades: a Dialogue Concerning Tar Water; Between Benjamin Smith, Lately Deceased, Dr. Hancock, and Dr. Garth, at Their Meeting upon the Banks of the River Styx* (London, 1744).

Anon., *Various Ironic and Serious Discourses on the Subject of Physick* (London, 1739).

Stephen Blankaart, *The Physical Dictionary*, 4th edn (London, 1702).

Blenheim Papers (British Library/London).

Charles Carter, *The Compleat City and Country Cook: or, Accomplish'd Housewife* (London, 1732).

Joseph Clutton, *A True and Candid Relation of the Good and Bad Effects of Joshua Ward's Pill and Drop. Exhibited in Sixty-Eight Cases* (London, 1736).

Corbyn Papers (Wellcome Library/London).

Diocese of London Papers (London Metropolitan Archives).

Alexander Gordon, *Lecture notes*. University of Aberdeen Special Collections (University of Aberdeen Library).

David Hartley, "An Account of the Contribution for Making Mrs Stephens's Medicines Public; with Some Reasons for it, and Answers to the most Remarkable Objections to it", *London Gazette* (26 Mar. 1738).

David Hartley, *Ten Cases of Persons Who Have Taken Mrs Stephens's Medicine for the Stone* (London, 1738).

Richard Lower, *Dr. Lower's, and Several Other Eminent Physicians Receipts*, 2nd edn (London, 1701).

John Moyle, *Chirurgus marinus: or, The Sea-Chirurgion*, 4th edn (London, 1702).

Edward Nourse, *Letter* (Royal Society/London).

Alexander Pope, "The First Epistle of the Second Book of Horace Imitated", in Aubrey Williams (ed.), *Poetry and Prose of Alexander Pope* (New York, 1969), pp. 241–54.

St Paul's Hammersmith Parish Records, 1766–84, in *Church of England Baptisms, Marriages, and Burials, 1538-1812*, accessed through ancestry.com.

David Turner, *The drop and pill of Mr. Ward, consider'd: In an epistle to Dr. James Jurin, Fellow of the College of Physicians. And of the Royal Society* (London, 1735).

Part Two
Ripples and Reflections

ACT UP New York. See also https://actupny.com; accessed 5 July 2023.

Ricardo Alonso-Zaldivar, "Trump Administration Wants to Import Cheaper Prescription Drugs from Abroad", *PBS Newshour* (19 Dec. 2019).

Mariana Alperin *et al.*, "The mysteries of menopause and urogynecologic health: clinical and scientific gaps", *Menopause* 26, n. 1 (2019), pp. 103–11.

Alzheimer Society of Canada, "What should Canadians know about aducanumab (a.k.a Aduhelm)?" See also https://alzheimer.ca/en/about-dementia/how-can-i-treat-dementia/what-aducanumab; accessed 10 Sept. 2023.

Anon. *A Generous Discovery of many Curious and Useful Medicines and Preparations; Both in Physic, Chymistry, Cookery, and Stiffenry* (London, 1725).

Association Canadienne de Thalidomide/Thalidomide Victims of Canada. See also https://thalidomide.ca/en; accessed 6 July 2023.

Fiona C. Baker *et al.*, "Sleep and sleep disorders in the menopausal transition", *Sleep Medicine Clinics* 13, n. 3 (2018), pp. 443–56.

Bayer Global Communications Department, "Bayer starts Phase III clinical development program OASIS with Elinzanetant". See also www.bayer.com/en/ca/bayer-starts-phase-iii-clinical-development-program-oasis-with-elinzanetant; accessed 10 Oct. 2023.

Pam Belluck and Rebecca Robbins, "Three F.D.A. Advisers Resign Over Agency's Approval of Alzheimer's Drug", *New York Times*, 10 June 2021 (updated 2 Sept. 2021).

Sanjoy Bhattacharya and Carlos Eduardo D'Avila Pereira Campani, "Reassessing the foundations: worldwide smallpox eradication, 1957–67", *Medical History* 64, n. 1 (2020), pp. 71–93.

Lorenzo Caputi *et al.*, "Missing enzymes in the biosynthesis of the anticancer drug vinblastine in Madagascar periwinkle", *Science* 360 (2018), pp. 1235–39.

Carter Center, "River Blindness Elimination Program". See also www.cartercenter.org/health/river_blindness/index.html; accessed 17 Aug. 2023.

Hugh Chamberlen, *A Proposal For the better Securing of Health* (London, 1689).

Bill Chappell, "3 Experts Have Resigned From An FDA Committee Over Alzheimer's Drug Approval", *National Public Radio* (11 June 2021).

ChemEurope Encyclopedia, "History of iodised salt". See also www. chemeurope.com/en/encyclopedia/History_of_iodised_salt. html#google_vignette; accessed 2 July 2023.

Jen Christensen and Betsy Klein, "Eli Lilly to cut insulin prices, cap costs at \$35 for many people with diabetes", *CNN* (1 Mar. 2023). See also www.cnn.com/2023/03/01/health/eli-lilly-insulin-prices-diabetes/ index.html; accessed 7 July 2023.

Elinor Cleghorn, *Unwell Women: Misdiagnosis and Myth in a Man-Made World* (Dutton, 2021).

Karen Codling *et al.*, "The legislative framework for salt iodization in Asia and the Pacific and its impact on programme implementation", *Public Health Nutrition* 20, n. 16 (2016), pp. 3008–18.

Committee on Oversight and Reform Staff Report, "Drug Pricing Investigation: Celgene and Bristol Myers Squibb, Revlimid", 30 Sept. 2020, United States House of Representatives.

Caroline Criado Perez, *Invisible Women: Data Bias in a World Designed for Men* (Vintage, 2019).

Anna Criddle, "Astellas' VEOZAH (fezolinetant) Approved by U.S. FDA for Treatment of Vasomotor Symptoms Due to Menopause", 13 May 2023. See also www.astellas.com/en/news/27756; accessed 10 Sept. 2023.

Stephanie Cross and Karen Watson, "Medical Hypothyroidism", *Healthline.com*. See also www.healthline.com/health/congenital-hypothyroidism#symptoms; accessed 2 July 2023

Ari Daniel, "Malaria is on the ropes in Bangladesh. But the parasite is punching back", *Goats and Soda: Stories of Life in a Changing World. National Public Radio* (20 Sept. 2023).

Ari Daniel, "A man dressed as a tsetse fly came to a soccer game. And he definitely had a goal", *Goats and Soda: Stories of Life in a Changing World. National Public Radio* (30 July 2023).

Hannah Devlin, "Drug for hot flushes will transform menopause treatment, doctors say", *Guardian* (20 May 2023).

Diabetes UK, "Who Invented Insulin?", *100 Years of Insulin*. See also https://www.diabetes.org.uk/our-research/about-our-research/our-impact/discovery-of-insulin; accessed 6 July 2023.

Selena Simmons Duffin and Carmel Wroth, "Maternal deaths in the U.S. spiked in 2021, CDC reports", *Morning Edition, National Public Radio* (16 Mar. 2023). See also www.npr.org/sections/health-shots/2023/03/16/1163786037/maternal-deaths-in-the-u-s-spiked-in-2021-cdc-reports; accessed 28 Sept. 2023.

Brian G.M. Durie *et al.*, "Bortezomib with lenalidomide and dexamethasone versus lenalidomide and dexamethasone alone in patients with newly diagnosed myeloma without intent for immediate autologous stem-cell transplant (SWOG S0777): a randomised, open-label, phase 3 trial", *Lancet* 389, n. 10068 (2017), pp. 519–27.

Katherine Eban, *Bottle of Lies: The Inside Story of the Generic Drug Boom* (HarperCollins, 2019).

Editorial, "Rapid drug access and scientific rigour: a delicate balance", *Lancet Neurology* 20, n. 1, (2021), P1.

Dean S. Elterman *et al.*, "The quality of life and economic burden of erectile dysfunction", *Research and Reports in Urology* 13 (2021), pp. 79–86.

Ezekiel J. Emanuel, *Which Country Has the Best Health Care?* (PublicAffairs, 2020).

Miriam Erick, "Frances Kathleen Oldham Kelsey", National Women's History Museum. See also www.womenshistory.org/education-resources/biographies/frances-kathleen-oldham-kelsey; accessed 21 Sept. 2023.

James Essinger and Sandra Koutzenko, *Frankie: How One Woman Prevented a Pharmaceutical Disaster* (History Press, 2018).

European Medicines Agency, "Aduhelm: withdrawal of the marketing authorisation application", 22 Apr. 2022. See also www.ema.europa.eu/en/medicines/human/withdrawn-applications/aduhelm; accessed 19 Sept. 2023.

European Medicines Agency, "From Laboratory to Patient: The Journey of a Medicine Assessed by EMA" (EMA, 2019).

European Medicines Agency, "How EMA Evaluates Medicines for Human Use". See also www.ema.europa.eu/en/about-us/what-we-do/authorisation-medicines/how-ema-evaluates-medicines#; accessed 20 Sept. 2023.

Harold Evans, "Foreword", in James Essinger and Sandra Koutzenko (eds), *Frankie: How One Woman Prevented a Pharmaceutical Disaster* (History Press, 2018).

Stephanie S. Faubion *et al.*, "Menopause symptoms on women in the workplace", *Mayo Clinic Proceedings* 98, n. 6 (2023), pp. 833–45.

C. Fauriant, "From bark to weed: the history of artemisinin", *Parasite* 18, n. 3 (2011), pp. 215–18.

Stephen Gandel, "Rep. Katie Porter gives Pharma executive the 'whiteboard' treatment", *MoneyWatch* (1 Oct. 2020).

Karen Geraghty, "Protecting the public: profile of Dr. Frances Oldham Kelsey", *Virtual Mentor: AMA Journal of Ethics* 3, n. 7 (2001), pp. 252–54.

Irwin Goldstein *et al.*, "The association of erectile dysfunction with productivity and absenteeism in eight countries globally", *International Journal of Clinical Practice* 73, n. 11 (2019), p. e13384.

April Grant, "FDA Approves Novel Drug to Treat Moderate to Severe Hot Flashes Caused by Menopause", US Food and Drug Administration (12 May 2023). See also www.fda.gov/news-events/press-announcements/fda-approves-novel-drug-treat-moderate-severe-hot-flashes-caused-menopause; accessed 30 Sept. 2023.

Matthew Harper, "Solving the Drug Patent Problem", *Forbes* (2 May 2002).

Robert Hart, "Roche is Discussing Alzheimer's Drug with FDA Following Regulator's Controversial Approval of Biogen's Aduhelm, CEO Says", *Forbes* (22 July 2021).

Harvard Health Publishing, "Cut salt – it won't affect your iodine intake" (1 June 2011). See also https://www.health.harvard.edu/heart-health/cut-salt-it-wont-affect-your-iodine-intake; accessed 25 Sep. 2023.

Harvard T.H. Chan School of Public Health, "The Nutrition Source: Iodine". See also www.hsph.harvard.edu/nutritionsource/iodine/;

accessed 25 Sept. 2023.

Harvard T.H. Chan School of Public Health, "The Nutrition Source: Salt and Sodium". See also www.hsph.harvard.edu/nutritionsource/salt-and-sodium/; accessed 25 Sept. 2023.

Adrienne Hatch-McChesney and Harris R. Lieberman, "Iodine and iodine deficiency: a comprehensive review of a re-emerging issue", *Nutrients* 14, n. 17 (2022), p. 3474.

Health and Human Services Press Office, "HHS Selects the First Drugs for Medicare Drug Price Negotiation" (29 Aug. 2023). See also https://www.hhs.gov/about/news/2023/08/29/hhs-selects-the-first-drugs-for-medicare-drug-price-negotiation.html; accessed 25 Sept. 2023.

Donna L. Hoyert, "Maternal mortality rates in the United States, 2021", National Center for Health Statistics Health E-Stats (2023); https://dx.doi.org/10.15620/cdc:124678.

John Innes Centre, "Madagascar periwinkle research uncovers pathway to cancer-fighting drugs", *ScienceDaily* (3 May 2018). See also www.sciencedaily.com/releases/2018/05/180503142809.htm; accessed 30 Sept. 2023.

Andrew Joseph, "3 Experts Have Resigned From An FDA Committee Over Alzheimer's Drug Approval", *STAT* (10 June 2021).

Kaiser Family Foundation Health Care Debt Survey, 25 February–20 March 2022.

Carliss Karasov, "Who reaps the benefits of biodiversity?" *Environmental Health Perspectives* 109, n. 12 (2001), pp. A582–87.

James H. Kim and Anthony R. Scialli, "Thalidomide: the tragedy of birth defects and the effective treatment of disease", *Toxicological Sciences* 122, n. 1 (2011), pp. 1–6.

Ashleigh Koss, "Update on Regulatory Submission for Aducanumab in the European Union", Biogen Company Statements (22 Apr. 2022). See also https://investors.biogen.com/news-releases/news-release-details/update-regulatory-submission-aducanumab-european-union-0; accessed 30 Sept. 2023.

Katherine Lang, "Progress and controversy in Alzheimer's research: aducanumab's FDA approval", *Medical News Today* (11 Jan. 2023), www.medicalnewstoday.com.

Linda Lear, "In Memoriam: Frances Oldham Kelsey". See also www. rachelcarson.org/frances-oldham-kelsey; accessed 21 Sept. 2023.

Lauren J. Lee *et al.*, "Increasing access to erectile dysfunction treatment via pharmacies to improve healthcare provider visits and quality of life: results from a prospective real-world observational study in the United Kingdom", *International Journal of Clinical Practice* 75 (2021), p. e13849.

Iliana C. Lega *et al.*, "A pragmatic approach to the management of menopause", *Canadian Medical Association Journal* 195 (2023), pp. e677–72.

Angela M. Leung, Lewis E. Braverman, and Elizabeth N. Pearce, "History of U.S. iodine fortification and supplementation", *Nutrients* 4 (2012), pp. 1740–46.

Gary F. Lewis and Patricia L. Brubaker, "The discovery of insulin revisited: lessons for the modern era", *Journal of Clinical Investigation* 131, n. 1 (2021), p. e142239; doi.org/10.1172/JCI142239.

Hilde Lindemann, "The woman question in medicine: an update", *The Hastings Center Report* 42, n. 3 (2012), pp. 38–45.

Mark S. Litwin, Robert J. Nied, and Nasreen Dhanani, "Health-related quality of life in men with erectile dysfunction", *Journal of General Internal Medicine* 13, n. 3 (1998), pp. 159–66.

Sydney Lupkin, "Drugmakers Blamed for Blocking Generics Have Jacked Up Prices and Cost U.S. Billions", *Kaiser Health News* (23 May 2018).

Soraya Machado de Jesus, Rafael Santos Santana, and Silvana Nair Leite, "Comparative analysis of the use and control of thalidomide in Brazil and different countries: is it possible to say there is safety?", *Expert Opinion on Drug Safety* 21, n. 1 (2022), pp. 67–81.

Howard Markel, "A grain of salt", *The Milbank Quarterly* 92 n. 3 (2014), pp. 407–12.

Alyson J. McGregor, *Sex Matters: How Male-Centric Medicine Endangers Women's Health and What We Can do About It* (Hachette Go, 2020).

Mitzzy F. Medellín-Luna *et al.*, "Medicinal plant extracts and their use as wound closure inducing agents", *Journal of Medicinal Food* 22, n. 5 (2019), pp. 435–43.

Charles Mégier, Grégoire Dumery, and Dominique Luton, "Iodine and thyroid maternal and fetal metabolism during pregnancy", *Metabolites* 13, n. 5 (2023), p. 633.

Giles Milton, *Nathaniel's Nutmeg; or, The True and Incredible Adventures of the Spice Trader Who Changed the Course of History* (Picador, 2015).

Kiho Miyazato, Hideaki Tahara, and Yoshihiro Hayakawa, "Antimetastatic effects of thalidomide by inducing the functional maturation of peripheral natural killer cells", *Cancer Science* 111, n. 8 (2020), pp. 2770–78.

Jennifer Rose V. Moleno, "The aducanumab controversy—how do physicians proceed?", Journal Watch, *New England Journal of Medicine* (29 Sept. 2021). See also www.jwatch.org/na54036/2021/09/29/aducanumab-controversy-how-do-clinicians-proceed; accessed 21 Sept. 2023.

Seeseei Molimau-Samasoni *et al.*, "Functional genomics and metabolomics advance the ethnobotany of the Samoan traditional medicine 'matalafi'", *Proceedings of the National Academy of Sciences* 118, n. 45 (2021), p. e2100880118.

K.M. Muraleedharan and M.A. Avery, "Therapeutic areas II: cancer, infectious diseases, inflammation & immunology and dermatology", in John B. Taylor and David J. Triggle (eds), *Comprehensive Medicinal Chemistry II* (Elsevier, 2007), pp. 765–814.

Edward R. Murrow, "See It Now", *CBS* (12 Apr. 1955).

S.J. Nass, G. Madhavan, and N.R. Augustine (eds), *Making Medicines Affordable: A National Imperative*. The National Academies of Sciences, Engineering, and Medicine – Health and Medicine Division; Board on Health Care Services; Committee on Ensuring Patient Access to Affordable Drug Therapies (Washington, DC, 2017).

National Institute of Child Health and Human Development, "How are Drugs Approved for Use in the United States?" See also www.nichd.nih.gov/health/topics/pharma/conditioninfo/approval; accessed 18 Sept. 2023.

National Library of Medicine, "Dr. Frances Oldham Kelsey" (2015). See also https://cfmedicine.nlm.nih.gov/physicians/biography_182.html; accessed 10 Sept. 2023.

Office of the Inspector General, "Delays in Confirmatory Trials for Drug Applications Granted FDA's Accelerated Approval Raise Concerns", OEI-01-21-00401, 29 Sept. 2022, US Department of Health and Human Services. See also https://oig.hhs.gov/oei/reports/OEI-01-21-00401.asp; accessed 21 Mar. 2024.

J. Orgiazzi and S.W. Spaulding, "Milestones in European Thyroidology: Jean-Francois Coindet (1774–1834)", European Thyroid Association. See also www.eurothyroid.com/about/met/coindet.html; accessed 5 Oct. 2023.

Davide Orsini and Mariano Martini, "Albert Bruce Sabin: the man who made the oral polio vaccine", *Emerging Infectious Diseases* 28, n. 3 (2022), pp. 743–46.

Elizabeth N. Pearce, "Is iodine deficiency reemerging in the United States?" *AACE Clinical Case Reports* 1, n. 1 (2015), p. e81.

Stephen Philips, "The power of no", *Medicine on the Midway* (2011), pp. 24–27.

Anna Poma *et al.*, "Anti-inflammatory properties of drugs from saffron crocus", *Anti-Inflammatory & Anti-Allergy Agents in Medicinal Chemistry* 11 (2012), pp. 37–51.

Seung Won Ra *et al.*, "The safety and efficacy of CKD-497 in patients with acute upper respiratory tract infection and bronchitis symptoms: a multicenter, double-blind, double-dummy, randomized, controlled, phase II clinical trial", *Journal of Thoracic Medicine* 13, no. 1 (2021), pp. 1–9.

Lisa Raffensperger, "How adding iodine to salt boosted Americans' IQ", *Discover Magazine* (23 July 2013).

Melissa Repko, "Walmart unveils low-price insulin as more patients with diabetes struggle to pay for drug", *CNBC* (29 June 2021).

Elizabeth Rosenthal, "KHN On NPR: The Uniquely American Problem of High Prescription Drug Costs: Interview with Elizabeth Rosenthal. By Scott Simon", *National Public Radio* (10 Jan. 2018).

Rita Rubin, "Collaboration and conflict: looking back at the 30-year history of the AIDS Clinical Trials Group", *JAMA* 314, n. 24 (2015), pp. 2604–06.

William Salmon, *A Rebuke to the Authors of the Blew-Book, Call'd The*

State of Physick in London (London, 1698).

Sarah Schulman, *Let the Record Show: A Political History of ACT UP New York, 1987–1993* (Farrar, Straus and Giroux, 2021).

Science Museum, "Thalidomide," Objects and Stories, 11 Dec. 2019. See also www.sciencemuseum.org.uk/objects-and-stories/medicine/thalidomide; accessed 21 Mar. 2021.

S. Singhal and J. Mehta, "Thalidomide in cancer: potential uses and limitations", *BioDrugs* 15, n. 3 (2001), pp. 163–72.

S.A. Skeaf, "Iodine and Cognitive Development", in David Benton (ed.), *Lifetime Nutritional Influences on Cognition, Behaviour and Psychiatric Illness* (Woodhead Publishing, 2011), pp. 109–28.

Peter Staley, "Antony Fauci Quietly Shocked Us All", *New York Times* (22 Dec. 2022).

A. Keith Stewart, "How thalidomide works against cancer", *Science* 343, n. 6168 (2014), pp. 256–57.

Katie Thomas, "The Story of Thalidomide in the U. S., Told Through Documents", *New York Times* (24 Mar. 2020).

Rosa Tikkanen *et al.*, "Maternal Mortality and Maternity Care in the United States Compared to 10 Other Developed Countries", Issue Briefs, The Commonwealth Fund (8 Nov. 2020). See also www.commonwealthfund.org/publications/issue-briefs/2020/nov/maternal-mortality-maternity-care-us-compared-10-countries; accessed 21 Mar. 2022.

Sarah Jane Tribble, "Drug Makers Play the Patent Game to Lock in Prices, Block Competitors", *Kaiser Health News* (2 Oct. 2018).

US Food and Drug Administration, "Drug Approval Process", Development and Approval Process: Drugs. See also www.fda.gov/drugs/development-approval-process-drugs; accessed 5 May 2023.

US Food and Drug Administration Approval Protocols, Code of Federal Regulations Title 21, vol 1, part 56: Institutional Review Boards. See also www.ecfr.gov/current/title-21; accessed 5 May 2023.

US Food and Drug Administration, "Exclusivity and Generic Drugs: What Does It Mean?". See also www.fda.gov/files/drugs/published/Exclusivity-and-Generic-Drugs--What-Does-It-Mean-.pdf; accessed 5 May 2023.

US Food and Drug Administration, "Frequently Asked Questions on Patents and Exclusivity". See also www.fda.gov/drugs/development-approval-process-drugs/frequently-asked-questions-patents-and-exclusivity; accessed 5 May 2023.

US Food and Drug Administration, "FDA Grants Accelerated Approval for Alzheimer's Drug", 7 June 2021. See also www.fda.gov/news-events/press-announcements/fda-grants-accelerated-approval-alzheimers-drug; accessed 5 May 2023.

Wulf H. Utian, "Psychosocial and socioeconomic burden of vasomotor symptoms in menopause: a comprehensive review", *Health and Quality of Life Outcomes* 3 (2005), article 47.

Gail A. Van Norman, "Drugs and devices: comparison of European and U.S. approval processes", *JACC: BTS* 1, n. 5 (2016), pp. 399–412.

Petra Verdonk, Elena Bendien, and Yolanbde Appelman, "Menopause and work: a narrative literature review about menopause and health", *Work* 72, n. 2 (2022), pp. 483–96.

Sheila C. Vir, "National iodine deficiency disorders control programme of India", in *Public Health Nutrition in Developing Countries* (Woodhead Publishing, 2011), pp. 575–605.

Jigang Wang *et al.*, "Artemisinin, the magic drug discovered from traditional Chinese medicine", *Engineering* 5 (2019), pp. 32–39.

Jennifer Whiteley *et al.*, "The impact of menopausal symptoms on quality of life, productivity, and economic outcomes", *Journal of Women's Health* 22, n. 11 (2013), pp. 983–90.

Brett Wilkins, "Katie Porter eviscerates Big Pharma CEO over 'exorbitant' drug prices", *Salon* (4 Oct. 2020).

World Health Organization, "Salt Reduction". See also www.who.int/news-room/fact-sheets/detail/salt-reduction; accessed 29 Sept. 2023.

Michael B. Zimmerman, "Research on iodine deficiency and goiter in the 19th and early 20th centuries", *Journal of Nutrition* 138 (2008), pp. 2060–63.

Acknowledgements

It is a great pleasure to be able to acknowledge and thank the many people and institutions who have made this work possible. The Portland Collection kindly gave permission to reproduce the Digby family portrait. I received funding and time from Seton Hall University and support from my colleagues in the English Department, who generously shouldered unpaid extra work while I was on sabbatical.

Thanks, thanks, and more thanks to my agent, Emma Bal at Madeleine Milburn, who is a book whisperer and a brilliant Virgil. Working with her is a privilege, thrill, and delight. I am deeply grateful to my editor Georgina Blackwell, Iain MacGregor, and the team at Head of Zeus for providing my work its first home, and to Eric Schmidt, my first editor at the University of California Press, Jyoti Arvey, and everyone at UCP for providing a second home on my own shore.

Librarians and archivists are among humanity's very finest exemplars; they have chosen a career of service, preserving human experience and helping people. Everyone at Walsh Library, present and retired, has made possible the last two decades of my research. I also am grateful to and for the librarians and staff at the Smithsonian's Dibner Library, London's Guildhall, the London Metropolitan Archives, the Society of Antiquaries of London, Lambeth Palace Library, the British Library's Western

Manuscripts and Rare Books divisions, the Wellcome Library, the National Archives at Kew, the Wiltshire and Swindon History Centre, the Warwickshire County Record Office, the Sheffield City Archives and Local Studies Library, the Aberdeen City and Aberdeenshire Archives, the Aberdeen Central Library, the Archives and Special Collections at the University of Glasgow, Tim Kirtley of Wadham College Library of Oxford University, Michelle Gait and the staff at the University of Aberdeen's Special Collections, Felix Lancashire and the staff at the Royal College of Physicians, Rupert Baker and the staff at the Royal Society, Darren Bevin and the staff of Chawton House and Chawton House Library, and the docents and staff of the Chelsea Physic Garden.

Although they might be surprised to find themselves here, I also wish to thank everyone at Scotrail who was on duty on 18 July 2023, one of the hottest days on record, when steel rails melted and electrical components burst into flame. I had to travel from Sheffield to Glasgow for research and saw first-hand their resilience and professionalism.

A veritable legion has been vital to this project, notably Lorna Dove, Lauren Kempf, and Marnie Doubek, as well as Marilyn Francus, Al Coppola, Laura Runge, Judith Zinsser, Anne Fernald, Donna Ritter, Laura Atwell, and Debi Rednik. Karen Brown Wheeler is quite simply the best friend ever. Lisa Thaler, development editor, helped me shape a stubbornly academic study into something for "real" people. My writing group (MG) and beta readers have been heroic. The other Visiting Residential Scholars at Chawton House Library – Emily Friedman, Victoria Joule, and Lindsay Seatter –Visiting Independent Scholar Mika Suzuki, and Summer Intern Christine Fulcher helped me articulate and test my ideas at the outset. It is impossible to measure how much I learned from Roger Gaskell and Caroline Duroselle-Melish and my classmates at the Rare Books School at the University of Virginia; I draw on it constantly. I am incredibly fortunate to work with Ros Ballaster, Jennifer Batt, Line Cottegnies, Leah Orr, Helen Wilcox, Aleksondra Hultquist, Gillian Wright, Claire Bowditch, Mel Evans, Margaret Ferguson, Maureen Bell, Paul Saltzman, and Mary Ann O'Donnell, and to have learned from the late Rob Hume. I also have been graced above my deserts with mentoring from three remarkable women: Jessica Munns, Martine Watson Brownley, and Elaine Hobby.

My family deserves all the gratitude that I can give and then some. Thanks to Amira and Tracy for their encouragement, sympathy, and terrific humour. A daughter-in-law could not possibly have better in-laws than Ed and Marilyn. For my parents, Susan and Steve, Amanda Foreman said it best in her biography of the Duchess of Devonshire: "I owe them much more than I can repay and am more grateful than my demeanor sometimes showed." David and Naama gave me the hearty kick I needed to begin this new journey; I never would have done it without them. To Stephen and our children, thank you. I love you.

Image Credits

Page 1
(Top left) Elizabeth Blackwell, *A Curious Herbal* (1739)
(Top right) Death as an apothecary's assistant making up medicines with a mortar and pestle for the apothecary attending a female patient who sits by the fireside. Watercolour by T. Rowlandson or one of his followers. Wellcome Collection
(Bottom) Lady Elizabeth Grey, Countess of Kent, *c.* 1619, Paul Van Somer *c.* 1577 or 1578 – *c.* 1621 or 22 / Tate Images

Page 2
(Top) Royal Collection of Physicians
(Bottom) The Portland Collection

Page 3
(Top) English Recipe Book, 17th century. Source: Wellcome Collection.
(Bottom left) Images from the History of Medicine (IHM) by Pierre Pomet, 1658–1699 / National Library of Medicine
(Bottom right) Dr [John] Huxham, English Medical Notebook / Wellcome Collection (photograph by Karen Bloom Gevirtz)

Page 4
(Top) Gift of Jill Spalding, 2022 / The Metropolitan Museum of Art
(Bottom) Jesus College, Oxford, Fellows Library

Page 5

(Top) The accomplished ladies rich closet of rarities / [J.S. (John Shirley)]. Wellcome Collection
(Bottom) A man and a woman demonstrating the process of fermentation and distillation in alchemy. Etching, *ca.* 17th century. Wellcome Collection

Page 6
(Top) A surgeon performing a lithotomy on a patient who is being restrained by three assistants, with five other anatomical illustrations. Engraving by J. Mynde. Wellcome Collection
(Bottom) Gift of Sarah Lazarus, 1891 / The Metropolitan Museum of Art

Page 7
(Top) An apothecary making up a prescription using scales, his wife holds a recipe for him and two assistants are working with the bellows and pestle and mortar. Line engraving by F. Baretta after P. Mainoto. Wellcome Collection
(Bottom left) Royal Pharmaceutical Society Museum
(Bottom right) © The Board of Trustees of the Science Museum

Page 8
(Top) Interior of a pharmaceutical laboratory with people at work; the shop is visible through a doorway. Engraving, 1747. Wellcome Collection
(Bottom left) Bettmann / Getty Images
(Bottom right) Busy Beaver Button Museum

Index

A

abbreviations and symbols,
109–10, 128–9, 185–6
*Accomplisht Lady's Delight
in Preserving, Physick,
Beautifying, and Cookery, The,*
87
Addison, Joseph, 156
aducanumab (Aduhelm), 245,
255–6
advertisements, for medication,
115–16, 119, 124–5, 196–7
AIDS Coalition To Unleash Power
(ACT UP), 254, 255
AIDS medication, 254–5
alchemy, 24, 28, 73, 109, 167–8,
181, 191
Alcorn, Richard, 139
Alles, Mark, 235
Allington, Charles, 170
Alzheimer's disease, 245
anatomy theatres, 151, 160
Anne, Queen of England, 27, 36,
116
Antimalarial medications, 258
antimony, 203, 210, 211, 212–13
apothecaries: 100–1, 124, 138,
216–17; and botany, 91–2;
and charity, 139–40, 141,
142–6, 149–50, 152; code,
45–6, 101–11, 185, 186;
community relationships, 148;
conflict with physicians, 27–8,
31, 95, 152, 154–5, 156–9;
history of, 15, 20–1, 27–8,
31; London guild laboratory,
150–1, 171; physic gardens,
94–8, 151; and prescription
system, 107–8, 125, 146; *vs.*
quack medication, 200–1,
202, 210, 219–20 (*See also*
Clutton, Joseph); and recipe
books, 45–6, 182, 184; shops,
218–9, 148; women as, 204,
220–1. See also Chelsea Physic
Garden; Worshipful Society of
Apothecaries
apothecary guilds: charity, 138,
139–40; establishment of,
27–8; medieval affiliations,
15, 21; physic gardens, 94–6;
revenue sources, 150–1
apothecary shops, 218–9
arsenic, 185, 210, 211–2
artemisinin, 257–8

V

Vesalius, Andreas, 23, 24–5, 26, 28, 268; *De Humani Corporis Fabrica Libri Septem*, 24–5, 26, 93
Viagra (sildenafil), 246, 247, 249
vinblastine, 260
vincristine, 260
viper wine, 60–3, 70

W

Wallis, Patrick, 111–12, 219
Walmart, 234
Ward, Joshua: background, 198–9; criticism of, 200–2, 203, 207–13, 215, 228–9; defamation suits, 205, 206, 211; defence of, 203–4, 205–6; performances of benevolence, 199–200; products and client base, 199; regulatory exception for, 229
Wasey, William, 144
Watts, James, 95
Welles, Benjamin, 116
Westminster Infirmary, 140–1, 142, 144
Whitaker, Tobias, *The Tree of Humane Life*, 80
Widowe's Treasure, The, 52
William III, King of England, 90, 161
Williams, Charles Hanbury, 200
Willis, Thomas, 189, 190; *Cerebri anatome*, 122; *Pathologiæ cerebri*, 122; *Pharmaceutice rationalis*, 61

Wilson, George, *A Compleat Course of Chymistry*, 180
Winter, Salvator, *A Pretious Treasury* (with Dickinson), 52
Wirtzung, Christopher, 50
Wise, Henry, 90
Wise, Prudence, 174, 183
witches and witchcraft, 16–17, 22, 62
Wolborough Feoffees and Widows' Charity, 140
woman of the house, as position, 37
Woman's Almanack, for the Year 1694, 87
women: as apothecaries, 204, 220, 221; bodies and experiences of, considered in medication discussions, 243–4, 246–7, 248–9, 255; as bonesetters, 213–15; domestic roles and household management, 16, 37–9, 71–2; exclusion from professional spaces and practice, 88–9, 107, 116, 125, 127, 172–3, 181–2, 221–7, 229–30; and gout, 116; in guilds, 108; in histories, 6, 262, 267–9; as physicians, 27, 100, 103–4, 125–6, 220–1, 228, 242–43, 257–8. *See also* domestic medicine; pregnancy and childbirth; shasthya kormi
Wood, Owen: *An Alphabetical Book of Physicall Secrets*, 52; *Choice and Profitable Secrets*